Fire Service Manual

Volume 3
Fire Safety

Basic Principles of Building Construction

HM Fire Service Inspectorate Publications Section
London: TSO

Published by TSO (The Stationery Office) and available from:

Online
www.tsoshop.co.uk

Mail, Telephone, Fax & E-mail
TSO
PO Box 29, Norwich, NR3 1GN
Telephone orders/General enquiries: 0870 600 5522
Fax orders: 0870 600 5533
E-mail: customer.services@tso.co.uk
Textphone 0870 240 3701

TSO Shops
123 Kingsway, London, WC2B 6PQ
020 7242 6393 Fax 020 7242 6394
16 Arthur Street, Belfast BT1 4GD
028 9023 8451 Fax 028 9023 5401
71 Lothian Road, Edinburgh EH3 9AZ
0870 606 5566 Fax 0870 606 5588

TSO@Blackwell and other Accredited Agents

ISBN 978-0-11-341189-4

Third impression 2007

This is a value added publication

Cover photograph: Douglas Specialist Contractors Ltd

Half-title page photograph: Douglas Specialist Contractors Ltd

Printed in Great Britain on material containing 75% post-consumer waste and 25% ECF pulp.

Printed in the United Kingdom for TSO

Basic Principles of Building Construction

Preface

This Volume of the Fire Service Manual replaces the previous Book 8. It also contains new sections dealing with fire safety generally.

In particular, the section (Chapter 1) dealing with The Regulation of Building Control recognises the significant impact of European Directives (The Construction Products Directive 89/106/EEC). The recent revision of Approved Document 'B' Fire Safety (2000 Edition) issued under the Building Regulations 1991 by The Department of the Environment, Transport and the Regions is also covered.

The European technical standards (Chapter 2) being introduced to support the Essential Requirements contained in the Directive are also included. The objective of harmonisation of building control throughout the 15 Member States of the European Union, means that fire officers will increasing be confronted with European Standards (Euro Norms (ENs)), rather than British Standards, and products that have been accorded the CE approval mark in their dealings with building proposals.

Chapter 1, therefore, provides an update of the present position in England and Wales, together with additional information applicable to Scotland. The impact of the European-wide activity is dealt with in some detail in Chapter 2.

Basic Principles of Building Construction

Contents

Basic Principles of Building Construction

Basic Principles of Building Construction

Introduction

Architects will usually design a building using those materials which they and their client agree will suit the required use and projected life of the building and remain within the financial restraints applied. An overriding factor is the relevant building controls that will be applied to the proposed building.

The building controls include certain standards that are applied to ensure that the behaviour of the materials used in the overall construction is acceptable in case of fire.

Firefighters must, therefore, try to be familiar with the properties of materials in so far as they could affect the safety of the public, themselves and the integrity of the building.

In recent years, this has become increasing difficult. Architects have developed novel uses of traditional building materials and new materials have been developed that allow a freedom of design unobtainable even a few years ago.

The move from prescriptive to functional building control has encouraged experimentation with both design and materials. The architect may use any material provided he can demonstrate its compliance with the prerequisite standards of fire behaviour. Such freedom brings problems for the firefighter who may be confronted by materials being used in unusual applications.

The characteristics of the individual components of a building structure will be affected by other components, especially in a fire. Firefighters are taught to detect signs of dangerous developments in a structure and they are encouraged to develop this knowledge by experience and observation.

This part deals with many types of building materials in use and gives examples of classification tests used in the United Kingdom.

Chapter 1 – The Regulation of Building Control

1.1 The Construction Products Directive (89/106/EEC)

With the current move towards harmonisation of building control within the European Union, the Essential Requirements of the Construction Products Directive will be met by harmonised methods of test that will be applicable to all Member States. At this present time, each Member State has its own system of building control based upon its National test methods. Such requirements inhibit the free movement of goods throughout the Community as materials tested, for example to BS476, will not be acceptable in another Member State who insists on classification according to its National test methods. Such barriers to the free movement of goods are not in keeping with the provisions of the Single European Act.

1.2 The European Committee for Standardisation (CEN)

The Commission (Enterprise Directorate-General) to produce the required standard methods of test has mandated the European Committee for Standardisation (CEN). The use of these CEN test methods will be mandatory in the UK and they will eventually replace the full range of test methods contained in BS476.

1.3 Fire Tests

Fire tests fall into two distinct categories, those tests intended to measure the resistance to fire of a material/product (fire resistance tests) and that intended to measure the fire behaviour of a material/product (reaction to fire tests).

All such tests are usually carried out in apparatus or rigs designed to simulate, as accurately as possible, the kind of situations the material under test could be involved in during a fire. The tests can be small (laboratory) scale involving specimens a few millimetres in size, through medium scale where the sample may be a metre or so in size up to full scale tests.

Unfortunately, there are an almost infinite variety of different fire situations which could involve any material and it is impossible to simulate them all, neither is it possible to always identify the contribution other components may make to a developing fire.

In a fire, such factors as:

(a) The severity of the fire;
(b) How long the material has been involved;
(c) The orientation/position of the material, e.g. wall, floor, ceiling, cladding or suspended;
(d) How the material is fixed into position;
(e) The reaction between the material and adjoining materials;
(f) The reaction of the material to the extinguishing medium used; and/or
(g) The standard of construction, workmanship etc.,

will have a bearing on the materials behaviour.

Any system of tests is designed to try and control the level of fire to an acceptable standard. The fact that materials used in the construction have satisfactorily attained a test requirement will not necessarily abolish the risk of fire or prevent it, it should, however, have a mitigating affect on fire development.

Primarily it is the contents of a building that give rise to the danger of fire and many test methods simulate the size of fire that could be predicted from the probable fire load within the compartment. Any assessment of the risk must take due account of the likely contents.

Architects will usually design a building using those materials which they and their client agree, will suit the required use and projected life of the building and remain within the financial restraints applied. An overriding factor is the relevant building controls that will be applied to the proposed building.

The building controls include certain standards that are applied to ensure that the behaviour of the materials used in the overall construction is acceptable in case of fire.

Firefighters must, therefore, try to be as familiar with the properties of materials in so far as they could affect the safety of the public, themselves and the integrity of the building.

In recent years, this has become increasing difficult. Architects have developed novel uses of traditional building materials and new or novel materials have been developed that allow a freedom of design unobtainable even a few years ago.

The move from prescriptive to functional building control has encouraged experimentation with both design and materials. The architect may use any material provided he can demonstrate its compliance with the prerequisite standards of fire behaviour. Such freedom brings problems for the firefighter who may be confronted with materials being used in unusual applications.

The characteristics of the individual components of a building structure will be affected by other components, especially in a fire. Firefighters should be able to detect signs of dangerous developments in a structure and they are encouraged to develop this knowledge by experience and observation.

1.4 The Building Control System in England and Wales

The regulation of Building Control in England and Wales is the responsibility of The Department for Transport, Local Government and the Regions (DTLR).

Regulations currently govern certain aspects of building design and construction in the interest of public health and safety, conservation of fuel and power, and making buildings accessible to disabled people. The regulation applies, in general, where building works are carried out, whether for new construction or in respect of alterations or extensions to existing buildings. It is important to note that they do not otherwise apply to buildings in use.

In 1991, under the provision of the Building Act 1984, the option was provided for Applicants to give, in some cases, a simple building notice instead of depositing full plans.

Approved Documents (see below) issued with the new regulations give technical guidance on how the requirements of the Regulations can be met. Builders and designers do not have to follow the Approved Documents if they can satisfy the local authority or Approved Inspector that their proposal meets the requirements of the Regulations. This is a more flexible system of control but it is one, which calls for the exercise of judgement by Building Control Officers in its interpretation and application.

1.5 The Fire Services Act 1947

Although the provisions of this Act are not intended to address specifically the question of fire safety in buildings at the stages of planning and construction, Section 1(1)(f) of the Act places a duty on the Fire Service to give advice.

As well as the requirement to consult under the Building Act, there are cases where Local Authority Building Control must consult the fire Authority under Section 16 of the Fire Precautions Act 1971. The Fire Services Act 1947 also obliges the Fire Authority to give advice on request on fire prevention and means of escape in case of fire.

1.6 Approved Inspectors (AIs)

Under the Building (Approved Inspectors etc.) Regulations 1985 (S1 1985 no 1066) (as amended) the Secretary of State can appoint suitably qualified "approved inspectors" who a developer/architect can employ to supervise all the aspects of the Building Regulations which would apply to a particular development or building.

The Approved Inspector is, in fact, a private building control officer (BCO) and is the equivalent of the local authority BCO; they may be individuals or bodies corporate. The Approved Inspector, like the BCO must consult with the fire authority in the usual way.

Under regulation 11 of the Building (Approved Inspectors etc.) Regulations 1985, an approved inspector has a duty to consult the fire authority in certain circumstances where it is proposed to erect a building.

The Departments recommend that an approved inspector should consider consultation in other circumstances, analogous to those in which a building control authority would consult the fire authority. This will reduce the risk of the approved inspector's client having to undertake further work to satisfy the fire authority, subsequent to the giving of a final certificate by the approved inspector [under the provisions of the Building Act and Approved Inspectors Regulations].

Therefore, unless otherwise specifically specified, reference to a building control authority within this document should be interpreted to mean approved inspector where appropriate.

The Regulations are administered by the Building Control Officers (BCOs) of local authorities who pass or reject plans, consider applications for relaxation or dispensation of provisions of the Regulations, and make inspections of the work during construction.

There is provision for supervision of building to be undertaken by Approved Inspectors instead of by the Local Authority Building Control. Besides the initial approval (or rejection) Local Authority Building Control Departments must also issue an informal letter of compliance and some even a formal Certificate at least for Part B which relates to fire. Approved Inspectors must issue a Certificate for all their projects.

1.7 The Approved Documents

The Secretary of State has approved a series of documents as practical guidance on meeting the requirements of Schedule 1 and regulation 7 of the Building Regulations 1991, as amended by the Building Regulations (Amendment) Regulations 1992 (SI 1992/1180), the Building Regulations (Amendment) Regulations 1994 (SI 1994/1850) and the Building Regulations (Amendment) Regulations 1995 (SI 1995/1356).

The detailed provisions contained in the Approved Documents are intended to provide guidance for some of the more common building situations. In some circumstances, alternative ways of demonstrating compliance with the requirements may be appropriate.

There is no obligation to adopt any particular solution in an Approved Document if the preference is to meet the relevant requirement in some other way.

Approved Document 'B' relates to fire safety, it comprises five parts:

B1 – Means of escape – that there is a satisfactory standard of means of escape for persons in the event of fire in a building.

B2 – Internal fire spread (linings) – that fire spread over the internal linings of buildings is inhibited.

B3 – Internal fire spread (structure) to ensure the stability of buildings in the event of fire; to ensure that there is a sufficient degree of fire separation within buildings and between adjoining buildings; and to inhibit the unseen spread of fire and smoke in concealed spaces in buildings.

B4 – External fire spread that external walls and roofs have adequate resistance to the spread of fire in the external envelope, and that spread of fire from one building to another is restricted.

B5 – Access and facilities for the fire service to ensure satisfactory access for fire appliances to buildings and the provision of facilities in buildings to assist firefighters in the saving of lives of people in and around the building.

Whilst guidance appropriate to each of these aspects is set out separately in Approved Document 'B', many of the provisions are closely interlinked. For example, there is a close link between the provisions for means of escape (B1) and those for the control of fire growth (B2), fire containment (B3), and the facilities for the fire service (B5). Similarly there are links between B3 and the provisions for controlling external fire spread (B4), and between B3 and B5. Interaction between these different requirements should be recognised where variations in the standards of provision are being considered. A higher standard under one of the requirements may be of benefit in respect of one or more of the other requirements. The guidance in the document as a whole should be considered as a package aimed at achieving an acceptable standard of fire safety.

A fire safety engineering approach that takes into account the total fire safety package can provide an alternative approach to fire safety. It may be the only viable way to achieve a satisfactory standard of fire safety in some large and complex buildings.

Some variation of the provisions set out in Approved Document 'B' may be appropriate where it applies to existing buildings, and particularly in buildings of special architectural or historic interest, where rigid compliance with the guidance might prove unduly restrictive. In such cases it would be appropriate to take into account a range of fire safety features, some of which are not addressed in any detail in Approved Document 'B', and to set these against an assessment of the hazard and risk peculiar to the particular case.

Factors that should be taken into account include:

(a) The anticipated risk of a fire occurring;
(b) The anticipated fire severity;
(c) The ability of the structure to resist the spread of fire and smoke;
(d) The consequential danger to people in and around the building.

A wide variety of measures could be considered and incorporated to a greater or lesser extent as appropriate in the circumstances. These include:

(a) The adequacy of means to prevent fire;
(b) Early fire warning by an automatic detection and warning system;
(c) The standard of means of escape;
(d) The provision of smoke control;
(e) Control of the rate of growth of a fire;
(f) The adequacy of the structure to resist the effects of fire;
(g) The degree of fire containment;
(h) Fire separation between buildings or parts of buildings;
(i) The standard of active measures for fire extinguishment or control;
(j) The facilities to assist the fire service;
(k) The availability of powers to require staff training in fire safety and fire routines e.g. Fire Precautions Act 1971 or registration or licensing procedures.
(l) Consideration of the availability of any continuing control under legislation that could ensure continued maintenance of such systems.

An alternative approach to the provision of fire safety in hospitals is available in Health Technical Memorandum (HTM) 81. Where guidance in that document is followed, Part B of the Building Regulations will be satisfied.

Approved Document

1.7.1 'B1' – Means of Escape

These provisions relate to building work and material changes of use which are subject to the functional requirements of B1, and they may therefore affect new or existing buildings. They are concerned with the measures necessary to ensure reasonable facilities for means of escape in case of fire. They are only concerned with structural fire precautions where these are necessary to safeguard escape routes.

They assume that in the design of a building, reliance would not be placed on external rescue by the fire service. The document has been prepared on the basis that, in an emergency, the occupants of any part of a building should be able to escape safely without any external assistance.

Special considerations however apply to some institutional buildings in which the principle of evacuation without assistance is not practical.

Attention is drawn to the fact that there may be other legislation imposing requirements for means of escape in case of fire with which the building must comply, and which operates when the building is brought into use. The main legislation in this area is the Fire Precautions Act 1971 where the Fire Precautions (Workplace) Regulations 1997 apply, generally enforced by the Fire Authority, which provides for the designation of certain uses of premises for which a fire certificate is required and, in the case of certain smaller premises, imposes a statutory duty on the occupiers to provide reasonable means of escape in case of fire.

Although there are no separate functional requirements under the building regulations to provide a means of giving warning in case of fire, the provision of an appropriate warning system is an essential element in the overall means of escape strategy for fire safety in occupied buildings.

In the Secretary of State's view the requirement of B1 will be met if:

(a) There are routes of sufficient number and capacity, which are suitably located to enable persons to escape to a place of safety in the event of fire;
(b) The routes are sufficiently protected from the effects of fire by enclosure where necessary;
(c) The routes are adequately lit;
(d) The exits are suitable signed; and if
(e) There are appropriate facilities to either limit the ingress of smoke to the escape route(s) or to restrict the fire and remove smoke.

All to an extent necessary depending on the use of the building, its size and height.

1.7.2 'B2' – Internal Fire spread (Linings)

The choice of materials for walls and ceilings can significantly affect the spread of a fire and its rate of growth, even though they are not likely to be the materials first ignited.

It is particularly important in circulation spaces where linings may offer the main means by which fire spreads, and where rapid spread is most likely to prevent occupants from escaping.

Two properties of lining materials that influence fire spread are the rate of spread of flame over the surface when it is subjected to intense heating, and the rate at which the lining materials gives off heat when burning. The guidance in this section provides for control of internal fire spread through control of these properties. This document does not give guidance on other properties such as the generation of smoke and fire.

In the Secretary of State's view the requirement of B2 will be met if the spread of flame over the internal linings of the building is restricted by making provision for them to have low rates of surface spread of flame, and in some cases to have a low rate of heat release, so as to limit the contribution that the fabric of the building makes to fire growth. The extent to which this is necessary is dependent on the location of the lining.

1.7.3 'B3' – Internal Fire Spread (Structure)

The fire resistance of an element of construction is a measure of its ability to withstand the effects of fire in one or more ways, as follows:

(a) Resistance to collapse, i.e. the ability to maintain loadbearing capacity (which applies to loadbearing elements only);
(b) Resistance to fire penetration, i.e. an ability to maintain the integrity of the elements (which applies to fire-separating elements);
(c) Resistance to the transfer of excessive heat, i.e. an ability to provide insulation from high temperatures (which applies to fire-separating elements).

'Elements of Structure' is the term applied to the main structural loadbearing elements, such as structural frames, floors and loadbearing walls. Compartment walls are treated as elements of structure although they are not necessarily loadbearing. Roofs, unless they serve the function of a floor, are not treated as elements of structure. External walls which as curtain wall or other forms of cladding which transmit only self weight and wind loads and do not transmit floor load are not regarded as loadbearing although they may need fire resistance to satisfy the requirement of B4.

In the Secretary of State's view the requirements of B3 will be met:

(a) If the loadbearing elements of structure of the building are capable of withstanding the effects of fire for an appropriate period without loss of stability;
(b) If the building is sub-divided by elements of fire-resisting construction into compartments;
(c) If any openings in fire-separating elements are suitably protected in order to maintain the fire integrity of the element; and
(d) If any hidden voids in the construction are sealed and subdivided to inhibit the unseen spread of fire and products of combustion, in order to reduce the risk of structural failure, and the spread of fire, in so far as they pose a threat to the safety of people in and around the building.

The extent to which any of these measures are necessary is dependent on the use of the building, and in some cases its size, and on the location of the element of construction.

1.7.4 'B4' – External Fire Spread

The provisions in this section limit the use near a boundary, of roof coverings which will not give adequate protection against the spread of fire over them. The term roof covering is used to describe constructions, which may consist of one or more layers of material, but does not refer to the structure as a whole. The provisions in this section are principally concerned with the performance of roofs when exposed to fire from the outside.

There are provisions concerning the fire properties of roofs in three other sections. In the guidance given to paragraph 5.3 there are provisions for roofs that are part of a means of escape. In the guidance to B2, there are provisions for the internal surfaces of roof lights as part of the internal lining of a room or circulation space. In the guidance to B3, there are provisions in Section 7 for roofs, which are used as a floor, and in Section 8 for roofs that pass over the top of a compartment wall.

1.7.5 'B5' – Access and Facilities for the Fire Service

The main factor determining the facilities needed to assist the fire service is the size of the building. Generally speaking firefighting is carried out within the building.

In deep basements and tall buildings firefighters will invariably work inside. They need special access facilities, equipped with fire mains. Fire appliances will need access to entry points near the fire mains.

In other buildings the combination of personnel access facilities offered by the normal means of escape, and the ability to work from ladders and appliances on the perimeter, will generally be adequate without special internal arrangements. Vehicle access may be needed to some or the entire perimeter, depending on the size of the building.

For dwellings and other small buildings, it is usually only necessary to ensure that the building is sufficiently close to a point accessible to fire brigade vehicles.

In taller blocks of flats, fire brigade personnel access facilities are needed within the building, although the high degree of compartmentation means that some simplification is possible compared to other tall buildings.

Products of combustion from basement fires tend to escape via stairways, making access difficult to fire service personnel. Providing vents can reduce the problem. Venting can improve visibility and reduce temperatures, making search, rescue and firefighting less difficult.

In the Secretary of State's view the requirement of B5 will be met:

(a) If there is sufficient means of external access to enable fire appliances to be brought near to the building for effective use;
(b) If there is sufficient means of access into, and within, the building for firefighting personnel to effect rescue and fight fire; and
(c) If the building is provided with sufficient internal fire mains and other facilities to assist firefighters in their tasks;
(d) If the building is provided with adequate means of venting heat and smoke from a fire in a basement.

These access arrangements and the facilities are only required in the interests of the health and safety of people in and around the building. The extent to which they are required will depend on the use and size of the building in so far as it affects the health and safety of those people.

This section should be read in conjunction with The Fire Precautions Act 1971 sections 13,14,15 and 16 and what is known as 'The Statutory Bar'.

1.8 Other Approved Documents

1.8.1 Basements for Dwellings

This document has been prepared by the Basement Development Group on behalf of the British Cement Association and the National House Building Council, and is approved by the Secretary of State under Section 6 of the Building Act 1984.

The document provides guidance on the design and construction of basements for dwellings including, in section 4, advice on fire safety.

1.9 Procedural Guidance Document

This document is intended to assist fire and building control authorities, approved inspectors, and other bodies or individuals, involved in the fire safety aspects of the Building Regulation approval processes.

The document:

(a) Gives guidance on the consultation processes between fire authorities, building control authorities and approved inspectors, necessary to reflect the statutory responsibilities placed upon the authorities, bodies or individuals involved;
(b) Gives guidance to ensure that the applicant receives clear, unambiguous advice at the earliest opportunity;
(c) Recognizes that the building control authority, or approved inspector, has the responsibility for ensuring that the building works concerned conform to the requirements of the Building Act and/or Building Regulations;
(d) Recognizes that the fire authority has a primary role in the building regulation approval process, in addition to either having a direct responsibility as an enforcing authority or as the primary advisor on fire safety to the controlling authority concerned.

In giving this guidance it is recognized that most fire and building authorities have sound arrangements in place and it is anticipated that authorities will build on those arrangements when considering these procedures.

1.9.1 Terminology

The following terms are used in the guidance document:

Applicant – the person seeking approval from a building control authority or employing an approved inspector to supervise building work. In practice the client for building work will often rely on an agent such as an architect to deal with the building control authority or approved inspector on his behalf. The term "applicant" is therefore used in this guide to include such agents. For convenience, "he" is used to include "she", as well as corporate persons.

Approved inspector – a person (whether an individual or a body corporate) approved under [section 49 of the Building Act 1984], for the purposes of supervision of building work (see above).

Building control authority – a local authority with duties under [section 91 of] the Building Act 1984 as regards application and enforcement of building regulations (a District or London Borough Council in England and a County or County Borough Council in Wales).

Building work – work of any of the sorts defined in regulation 3(1) of the Building Regulations 1991 (SI 1991 No 2768). This includes:

(a) Erection or extension of a building;
(b) Provision or extension of a controlled service or fitting;
(c) Material alteration of a building or controlled service or fitting;
(d) Work required under the building regulations in connection with a material change of use.

The expressions "controlled service or fitting" and "material alteration" and "material change of use" are defined in the Building Regulations 1991.

The guidance contained in this document reflects the roles and responsibilities placed on both the fire and building control authorities, in ensuring that appropriate fire safety provisions are included in the design and construction of buildings.

The guidance recognizes that:

(a) The building control authority has the primary responsibility for ensuring that any building works comply with the requirements of the Building Regulations;
(b) The fire authority has a number of roles in ensuring compliance or giving advice in respect of fire safety. These are:
    (i) Where appropriate the primary responsibility as an enforcing authority,
    (ii) Where appropriate the function, identified within the legislation concerned, of advising the authority with ultimate enforcing/controlling responsibility,
    (iii) When requested giving advice in respect of fire safety matters.

The guidance in this document is intended to avoid an occupier of a building, approved under Building Regulations, having to carry out further structural or other alterations to meet the needs of fire safety legislation applying to its first use. To assist in this regard The Fire Precautions Act 1971 contains a **"statutory bar"** that prohibits the fire authority from requiring any additional building works under that legislation, other than in those instances where the circumstances concerned were not known at the time of the building regulation application and subsequent approval.

It is important that any consultation process recognizes that, other than in those limited circumstances specifically identified under the Fire Precautions Act 1971, it is the authority which has the final responsibility for licensing/registering/controlling the premises when occupied that determines whether the fire precautions proposed are adequate for the building when in use.

It is therefore important for all relevant authorities to be consulted on the fire safety standards being proposed and for the views of the controlling authority, or the authority acting on their behalf, to be taken into account and, if at all possible, incorporated into the final approval of the building works.

The following guidance is intended to ensure as far as practical a common approach on the consultation arrangements irrespective of the legislation that will apply to the building for its first use(s).

The building control authorities will usually become aware of proposed building works via:

(a) Requests for advice;
(b) Formal applications for approval;
(c) Building Notices; or
(d) Consultation by the fire authority.

The fire authorities will usually become aware of proposed building works via:-

(a) Requests for advice;
(b) Formal applications for approval in respect of existing buildings;
(c) Formal notification/consultation by applicants or authorities;
(d) Re-inspections of existing buildings; or
(e) Consultation by the building control authority.

The guidance on consultation arrangements is therefore structured as follows:

(a) New buildings or unoccupied buildings (Section 2),
(b) Occupied buildings (Section 3),
(c) Deemed to consult procedures (section 4),
(d) Other consultation (Section 5),
(e) Requests for advice (Section 6),
(f) Exchange of information (Section 7), and
(g) Informal disputes procedure (Section 8).

The consultation process between fire and building control authorities must recognize the time constraints, imposed by the various legislation concerned, on both authorities. However it must also take into account the need for the recipient authority to have adequate time to consider the proposals and prepare a response. Responses to consultation between authorities should be within

14 days of the receipt of the consultation documentation and, therefore, the consulting authority should ensure that it takes this period into account in its administrative procedures.

1.10 Consultation arrangements – new buildings or unoccupied buildings

When the building authority receive a valid application for Building Regulation approval in respect of buildings for which either the first use will be designated under the Fire Precautions Act, the Fire Precautions (Workplace) Regulations, as amended, or the building is one that may require a licence, registration or other statutory approval, they should consult with the fire authority.

If the building authority receive an application for a complex building project or one that involves the use of fire safety engineering techniques, and they consider it appropriate to set up a meeting to discuss the proposal, they should ensure that the fire authority is represented at the meeting.

The consultation should include sufficient plans and detail of the application to enable the fire authority to consider the application in respect of its intended first use.

The building authority should indicate whether or not they intend to approve the application and include a draft of any conditions of approval they are considering. If they are minded to reject the application, they should notify the fire authority that they have received and rejected the application.

Where the fire authority are satisfied with the application, they should inform the building authority of that fact in writing. If appropriate, they should return one copy of the plan, dated and marked with their approval.

If the fire authority is dissatisfied with an application, they should arrange an early meeting with the building authority to discuss their concerns. Alternatively, they should respond in writing stating the reasons for their dissatisfaction, which may include a marked up plan showing the changes they would wish to be incorporated in the proposal.

Any response from the fire authority should, where appropriate, be clearly identified as to:

(a) Those fire safety provisions necessary, in the opinion of the authority, to comply with the Building Regulations;
(b) Those fire safety provisions necessary, in the opinion of the fire authority, to meet any other legislative responsibilities of the applicant;
(c) Fire safety provisions, in addition to those referred to above, which it is recommended should be provided, together with the reasons for such recommendations; and
(d) Any fire safety provisions, proposed by the applicant, which in the opinion of the fire authority would not be necessary to meet legislative responsibilities.

Where the building authority agree with the fire authority's comments in respect of those aspects relating to the Building Regulations, they should either reject the application or give a conditional approval including the changes the fire authority wish to see incorporated in the proposal.

Where the fire authority has made other comments on matters needed to meet the needs of other legislation, or recommendations, the building authority should pass these comments to the applicant, clearly identifying the source of the comments and that they are made in addition to any works that may be necessary to satisfy the requirements of the Building Regulations.

1.11 Failure of authorities to agree building regulation issues

Where both authorities cannot agree those issues necessary to meet the requirements of the Building Regulations, the matter should be referred to the dispute procedure (See section 8 of the Procedural Guidance Document).

1.12 Statutory consultation arrangements – occupied buildings

One of the major issues to be taken into account when considering proposals for building works in existing occupied buildings is the potential effect of the proposed works on the existing fire safety arrangements, both during the construction phase and in the completed scheme.

Taking the above into account a duty is placed upon occupiers, under the Fire Precautions Act and other legislation, to obtain the approval of the controlling authority before commencing any alterations to existing occupied premises.

The following guidance recognizes that, on occasions, occupiers may fail to meet the obligations outlined above. Therefore whilst the guidance given is intended to provide a practical procedure for ensuring that adequate fire safety arrangements are provided, it is not to be considered as an alternative to, or defence against failing to meet, the formal responsibilities imposed by the relevant legislation.

The fire authority, in the case of premises subject to the Fire Precautions Act, has a direct responsibility to ensure that the overall standard of fire safety throughout the building is maintained throughout any construction phase.

1.13 Initiation of notices by building authority

Where a building authority receives an application for alterations or extensions to an occupied building in which the fire authority has either a direct or indirect interest in the fire safety arrangements, the application should be dealt with as detailed in Section 2.

In responding to a consultation in respect of an occupied building the fire authority may give directions (in case of premises subject to the Fire Precautions Act), or advice (in the case of other premises), as to how the proposed works should be carried out, both to preserve the fire precautions in the building whilst the work is in progress and to meet their statutory responsibilities under the Construction (Health, Safety and Welfare) Regulations 1996. Any such directions or advice must be clearly identified as outlined below and subsequently included with any notice or approval to the applicant.

1.14 Initiation of notices by fire authority

The fire authority may consider serving a notice to initiate alterations to the fire safety arrangements of the building for one of two reasons:

(a) When the fire authority receive notification from the applicant of proposed alterations involving building work, proposed changes of use or other alterations,
(b) Following an inspection of the premises to monitor the maintenance of existing fire safety arrangements, and as a result of identifying alterations to the premises, processes or other issues that they consider affect the adequacy of the fire safety arrangements.

When the fire authority receive notification from the applicant of proposed alterations involving building work, they should in their acknowledgment of the application advise the applicant to submit an application to the building authority for Building Regulation approval before work is commenced.

The fire authority should consider the notification, or the alterations to the building, and determine as to whether in their opinion the proposals or alterations are satisfactory.

If the fire authority consider that the proposals are unsatisfactory they should prepare a notice of:

(a) Those fire safety provisions necessary, in the opinion of the authority, to comply with the Building Regulations;
(b) Those fire safety provisions necessary, in the opinion of the fire authority, to meet any other legislative responsibilities of the applicant;
(c) Fire safety provisions, in addition to those referred to above, which it is recommended should be provided, together with the reasons for such recommendations;
(d) Work necessary to maintain fire safety in the altered building;
(e) Any directions they consider necessary as to how the work should be progressed; and
(f) Any fire safety provisions, proposed by the applicant, which in the opinion of the fire authority would not be necessary to meet legislative responsibilities.

Before confirming acceptance of any proposal received or serving a notice, the fire authority should consult with the building authority. The consultation should include a copy of the notification from the occupier and any plans accompanying the notice (in those cases where notification was received), details of the use of the premises concerned, a copy of the notice they intend to serve and any appropriate plans. Where the notice is in respect of the outcome of a fire risk assessment, details of that outcome should be included.

Where the building authority are satisfied with an application or the notice prepared by the fire authority, they should inform the fire authority of that fact in writing. If appropriate, they should return one copy of the plan, dated and marked with their approval.

If the building authority are dissatisfied with an application, or those aspects of the notice prepared by the fire authority that in their opinion relate to the Building Regulations, they should arrange an early meeting with the fire authority to discuss their concerns. Alternatively, they should respond in writing stating the reasons for their dissatisfaction, which may include a marked up plan showing the changes they would wish to be incorporated in the proposal.

The fire authority should ensure that any formal notice to the applicant is amended to include the building authority's comments, where necessary under a separate heading.

1.15 Simultaneous application to both authorities

If at any time during the consultation period, the building authority receive an application for Building Regulation approval in respect of the proposed alterations, the application should be dealt with in accordance with the procedures in Section 2, except that the contents of any response from the fire authority should be as detailed in paragraph 3.10.

1.16 Building authority response to consultation by fire authority

When responding to a consultation by the fire authority and where no separate application for a building regulations application has been received, the building authority should notify the fire authority of any additional fire safety provisions which, in their opinion, are necessary for the proposed building works to receive Building Regulation approval. The fire authority should ensure the comments of the building authority are incorporated in the notice, where necessary under a separate heading.

1.17 Application for building regulation approval following issue of notice by fire authority

When considering an application for Building Regulation approval resulting from a notice served by the fire authority, the building authority should have regard to the previous consultation. If they are satisfied that the application incorporates those matters previously notified to or agreed with the fire authority and can approve the application unconditionally, they may invoke the "deemed to consult" procedures which may be found in Section 1.4 of this guide.

1.18 Building authority rejection of building regulation application following issue of notice by fire authority

If the building authority are minded to reject an application resulting from a notice from the fire authority which has been the subject of previous consultation, they should immediately consult the fire authority and advise them of their concerns.

1.19 Failure of authorities to agree building regulation issues

Where both authorities cannot agree those issues necessary to meet the requirements of the Building Regulations, the matter should be referred to the dispute procedure.

1.20 Building Notice

By definition – "A notice in prescribed form given to the local authority under regulations 11(1)(a) and 12 informing the authority of proposed building works."

A major procedural innovation introduced by the 1985 Regulations and now to be found in the 1991 Regulations, is based on service of a building notice. There is no approval of plans.

A person intending to carry out building work or make material change in the use of a building may give a building notice to the local authority unless the building concerned is one which is designated under the Fire Precautions Act 1971.

There is no prescribed form of building notice. The notice must be signed by the person intending to carry out the work or on his behalf, and must contain or be accompanied *inter alia* by the following information:

(a) The name and address of the person intending to carry out the work;
(b) A statement that it is given in accordance with regulation 11(1)(a);
(c) A description of the location of the building to which the proposal relates and the use or intended use of the building;
(d) If it relates to the erection or extension of a building it must be supported by a plan to a scale not less that 1 : 1250, showing size and position in relation to streets and adjoining buildings on the same site, the number of storeys, details of the drainage and the precautions to be taken in building over any drain, sewer or disposal main. Where local legislation applies, the notice must state how it will be complied with.

The local authority are not required to approve or reject the building notice and, indeed, have no power to do so. However, they are entitled to ask for any plans they think are necessary to enable them to discharge their building control functions and may specify a time limit for their provision. They may also request the person giving the notice for information in connection with their linked powers under sections 18 to 21 of the Building Act 1984.

A building notice remains in effect for a period of three years from the date on which it was given to the local authority. If the work has not been commenced in that period the building notice lapses automatically.

1.21 Completion Certificate

The local authority must issue a completion certificate when they are satisfied, after having taken all reasonable steps, that the Schedule 1 requirements are met. Its issue is mandatory in respect of fire safety requirements, i.e. where the building is to be put to a designated use under the Fire Precautions Act 1971, but in other cases the authority need issue the certificate only where they have been requested to do so.

The completion certificate is evidence – but not *conclusive* evidence that the requirements specified in the certificate have been complied with.

1.22 The Building Control System in Scotland

The legislation, which controls building construction and building operations in Scotland, is the Building (Scotland) Act 1959, as revised.

The building control system is based on the administration, by local authorities of the Building Standards (Scotland) Regulations 1990, as amended. These Regulations apply to the construction, alteration, extension or demolition of a building or part of a building or to any change of use which attracts additional or more onerous requirements.

The essential purpose is to safeguard people in and around buildings. It is a pre-emptive system, ensuring as far as possible that proposed buildings will not contravene the Regulations and that on completion; the buildings do in fact comply.

The main technical requirements are now in the form of short functional statements (Regulations 10–33). These are supported by a detailed set of Technical Standards, which are, in effect, Mandatory as a result of Regulation 9. The effect of Regulation 9 is that there are basically three methods by which the requirements of Regulations 10–33 can be satisfied:

(a) By compliance with the relevant standards set out in the supporting Technical Standards;
(b) By conforming to the provisions which are stated in the Technical Standards to be 'deemed to satisfy the relevant standards';
(c) By any other means which can be shown to satisfy the relevant standards.

All building construction must comply with the standards prescribed by the Building Standards

Regulations and it is an offence for anyone to carry out building work, other than that specifically exempted, without a building warrant having been obtained from the local building control authority, who will advise on what drawings and other information are required to enable them to check for compliance with the Regulations.

During construction, building control staff may inspect the works, and the local authority must be notified when building begins and subsequently at specified stages. Tests may be required to establish compliance with the Regulations.

Before the building may be occupied it is necessary to obtain from the local authority, a certificate of completion. This provides formal confirmation that the building has been erected in accordance with the warrant and with the Building Standards Regulation.

There are arrangements for the local authority to permit relaxation from the requirements of the Regulations. In these, the key consideration is whether it is unreasonable in the particular circumstances for the specific requirements to be applied. If a relaxation application concerns a departure from Regulation 13 (means of escape from fire etc.) the local authority must consult the fire authority.

There is no statutory requirement for the building control authority to consult the fire authority about applications where no relaxation is involved. This is not deemed to be necessary as brigades are represented on the Building Standards Advisory Council which advises the Secretary of State on the form and content of the Regulations and Technical Standards. Many brigades, however, have developed effective local arrangements for informal consultations on a day-to-day basis.

1.23 The Technical Standards

The Technical Standards book includes most of the Building Standards (Scotland) Regulations 1990 as well as the relevant standards and the 'deemed to satisfy' provisions together with general information on the building control system in Scotland. It should be noted that the Technical Standards must be applied as a whole and in many instances the standards, in several parts, act in concert.

The parts of the Technical Standards, which are of particular relevance to the fire officer, are:

1.23.1 Part A – General

The intention of Part A is to give general information, which is essential to the application of the Technical Standards. It also gives a complete extract of those regulations not covered in the individual parts of the Standards.

1.23.2 Part D – Structural Fire Precautions

The intention of this Part is to ensure that the structure of a building and the parts of a building will remain stable in the event of fire and will restrict the spread of fire and smoke within the building and spread of fire between buildings.

To restrict the internal spread of fire, a building may have to be divided into compartments separated from each other by compartment walls or compartment floors intended to provide a complete barrier between the compartments. In practice, the continuity of such walls and floors will have to be breached by openings for circulation or service and special precautions are necessary where this occurs to maintain the effectiveness of the barrier.

The acceptable size of a compartment for this purpose is determined by its likely fire load which is, in turn, influenced by the purpose group of the building, in which it is situated and the provision, or otherwise, of active fire protection measures such as a sprinkler installation.

The elements of structure must continue to function, and remain capable of supporting and retaining any necessary protection of escape routes and fire access routes during a fire, for an adequate period of time.

In order to reduce the danger to the occupants of other buildings, suitable separation must be provided between one building and another by either structure or distance. The acceptable distance between a building and its relevant boundary is dictated by the amount of heat likely to be radiated in the event of fire by the wall facing that boundary.

The extent of openings, or other unprotected areas in the wall and the likely fire load of the building will influence this. Acceptable methods of calculating separation by distance are given in the 'Provisions deemed to satisfy the standards'. Provision is also made to reduce the likelihood of fire spreading to roofs from an external source.

Information on methods of determining fire resistance, non-combustibility and external fire exposure designation of roofs is contained in the Appendix to Part D.

1.23.3 Part E – Means of escape from fire facilities for firefighting and means of warning of fire in dwellings.

The intention of this part is to provide all users of a building with adequate means of escape from fire. It also requires the provision of certain fixed firefighting equipment, means of access for firefighting, and means of warning of fire in a dwelling.

The intention of the requirement for means of escape is that everyone within a building may reach either a place of safety or, in certain circumstances, a protected zone within a reasonable travel distance. The requirements for the number and widths of exits assume a unit width of 530mm per person and a rate of discharge of 40 persons per minute. The number of escape routes is determined by:

(a) The purpose group of the building;
(b) The occupancy capacity serviced;
(c) The height of the storey above ground or its depth below ground; and
(d) The travel distance involved.

A building must be planned so that:

(a) Every escape route leads to a place of safety;
(b) Every stair or ramp which forms part of an escape route, except an escape stair from a gallery, is protected from fire, from smoke and hot gases which might obscure or obstruct the escape route and, in higher buildings, from the effects of weather. In certain cases, provision is made within a protected zone for a refuge for people with a disability.
(c) Within those parts of a building where people are at greatest risk, the layout of the building is such as to limit that risk to the utmost practical extent; and
(d) In certain residential buildings, which have only one escape route, provision is also made for rescue by way of a suitable window.

Suitable provision must be made for access to the outside of a building for firefighting and rescue vehicles from a public road. A water supply installation must be available and, in the case of high buildings, suitable provision must be made for firefighting within the building.

1.23.4 Part F – Facilities for people with a disability

The intention of this part is to require that a building, to which it applies, is accessible to people with a disability and is provided with aids to assist those with impaired hearing and those with impaired sight to move around the building. Fire escape requirements from storeys accessible to people with a disability will be found in Part E.

1.24 The Building Control System in Northern Ireland

There is not parity at the moment between the regulations in England and Wales and the regulations in Northern Ireland although this is planned to be achieved in the future.

The Regulations currently in force are The Building Regulations (Northern Ireland) 1994. The advice on fire safety aspects is found in Technical Booklet 'E'.

1.24.1 Technical Booklet 'E'

This Technical Booklet has been prepared by the Department of the Environment for Northern Ireland and provides for certain methods and standards of building, which, if followed, will satisfy the requirements of the Building Regulations (Northern Ireland) 1994.

There is no obligation to follow the methods or comply with the standards set out in the Technical Booklet.

This Technical Booklet relates only to the requirements of regulations E2, E4, E6, E8 and E10 and is in six sections:

†1 Means of Escape
†2 Internal fire spread – Linings
†3 Internal fire spread – Structure
†4 External fire spread
†5 Facilities and access for the fire brigade
†6 Common provisions.

The provisions set out in the Technical Booklet under Sections 1 to 5 deal with different aspects of fire safety. Whilst the provisions appropriate to each of these aspects are set out separately, many of the provisions are closely interlinked. For example, there is a close link between the provisions of means of escape (†1) and those for the control of fire growth (†2), fire containment (†3) and facilities for the fire brigade (†5). Similarly, there are links between section 3 and the provisions for controlling external fire spread in section 4 and between section 3 and section 5.

Interaction between these different provisions should be recognized when considering alternative solutions as the adaptation of a higher standard of provision in respect of one aspect may be of benefit in respect of the provisions relating to one or more other aspects. Thus, the provisions in the Technical Booklet as a whole should be considered as a package aimed at achieving an acceptable standard of fire safety.

Chapter
2

Chapter 2 – European Directives and The Standardisation Process

There is a direct relationship between European Directives and Standardisation, particularly in the field of fire safety. The principal Directive, The Construction Products Directive (see below) contains Essential Requirements, which must be met before products can be placed on the open European market.

The principal means of demonstrating compliance and therefore achieving the ***CE*** mark is to use harmonised European Standards.

2.1 Why Standardisation is important

The importance of international standardisation to the elimination of barriers to European and global trade is reflected in the marked swing in Standards activities from National to European and international standards production. European and International standards work within British Standards has grown from about 11% of 9,000 work items in 1983 to about 93% of 17,600 work items today.

2.2 The British Standards Institution (BSI)

The British Standards Institution is the independent national body in the United Kingdom responsible for preparing and promoting British Standards. It provides the gateway for UK representation in European and International standards organisations.

It is a non-profit distributing organisation formed by subscribing members and committee members and incorporated by Royal Charter since 1929. The Institution derives its unique role as the national standards body from its Charter, its membership of the International and European standards organisations and its receipt of some government funding for standards work. Other BSI activities include testing and certification of products.

Standardisation within the Health and Environment area (which includes fire safety) is conducted in three fora:

(a) National;
(b) European; and
(c) International.

2.3 National Standards

National standards are drafted by experts and agreed by committees of representatives (technical committees and sub committees) drawn from industry, government, consumers, users, professionals etc. A chairman manages the work of these committees, supported by a secretary who is usually a BSI staff member, but occasionally is a representative of an industrial organisation.

2.4 European Standards

With the recognition of the importance of open markets within Europe an increasing number of standards are produced at European level by two bodies:

(a) European Committee for Standardisation (**CEN**); and
(b) European Committee for Electrotechnical Standardisation (**CENELEC**).

A committee of representatives also drafts these standards. The members of these committees are drawn from the committee members of the various national standards bodies, which form the European standards bodies.

By common agreement, all the national standards bodies adopt the European standards as their own national standards and withdraw any existing national standards, which conflict with them. It is crucial, therefore, that the UK is actively involved in European standards work. BSI is responsible for organising the UK input into all European standards.

2.5 International Standards

In the interests of global trading, standards should be applicable not only in Europe but also across the whole world. Two bodies undertake international standardisation:

(a) International Organization for Standardisation (**ISO**); and
(b) International Electrotechnical Commission (**IEC**).

The membership of these bodies comprises the national standards bodies of more than 90 countries around the world. They operate on a global scale in a similar way to the European standards bodies. If possible, European standards are adoptions of International standards, or are proposed as International standards, in order to avoid duplication of effort or conflicting standards at European or International level.

There are no requirements or agreement for national standards bodies to adopt International standards or to withdraw conflicting National standards, but it is still important that the UK involves itself in much of the work of ISO and IEC, particularly if it relates to an area where there is no European activity.

The International Organization for Standardisation is a worldwide federation of national standards bodies (ISO member bodies). The work of preparing International Standards is normally carried out through ISO technical committees. Each member body interested in a subject for which a technical committee has been established has the right to be represented on that committee. International organizations, governmental and non-governmental, in liaison with ISO, also take part in the work. ISO collaborates closely with the Electrotechnical Commission (IEC) on all matters of Electrotechnical Standardisation.

2.6 National Standards implementing European Standards

Since the 1970s, BSI has published International (ISO) and European (EN) standards as identical British Standards, with a national foreword. Such standards are easily distinguishable, as they will carry the prefix of BS ISOxxx, or BS Enxxxx. In the true spirit of the Vienna Agreement (see below) and international harmonisation, it is increasing common to find technical standards that have achieved harmonisation in all three organisations, i.e. BS ISO EN****.

In Europe, national implementation is a requirement under CEN/CENELEC Common Rules, following adoption, by weighted majority, of a European Standard (EuroNorm – EN). It is important to remember that the action of the national standards body cannot make a European standard compulsory. Any British Standard becomes legally binding only by contract, or if compliance is claimed, or if Legislation makes it mandatory.

The obligation of CEN members to implement European standards by giving it the status of a National standard means that the technical content of an EN is presented in identical form in each country and is of equal validity. In addition, CEN/CENELEC can impose a 'standstill' agreement, which prevents publication of any divergent National standard whilst European work is in hand.

2.7 The Vienna Agreement

With two separate international standards bodies (ISO and CEN) there is a real need to ensure there is no duplication of effort. It would be pointless exercise to produce two standards covering the same area of work and to avoid this, the two standards bodies have signed a formal agreement to keep each other appraised of their work. The agreement was drawn up at a meeting in Vienna and is widely known as 'The Vienna Agreement'.

Therefore, if standards are being developed within ISO and similar standards are required under a Mandate for a Directive, CEN will agree to adopt the ISO Draft when it is available and it will be submitted for 'dual' voting procedures, i.e. it will be circulated throughout both the ISO and CEN membership for acceptance.

2.8 Standards and European Union Legislation

2.8.1 A standard is:

'A document, established by consensus and approved by a recognised body, that provides, for common and repeated use, rules, guidelines or characteristics for activities or their results, aimed at the achievement of the optimum degree of order in a given context.'

2.8.2 The 'Old' Approach

Since the 1960s, Directives issued under Article 100 of the Treaty of Rome established free circulation objectives for Member States, linked to detailed annexes setting out technical requirements. Keeping these annexes up-to-date proved to be an unexpected burden, the flow of new Directives dried up and the Commission revised its policy in the early 1980s.

2.8.3 The 'New' Approach

Directives following the 1985 New Approach to technical harmonisation and standards are based on the new Article 100a introduced by the Single European Act.

These Directives promulgate mandatory essential requirements as a common basis for national Regulations and look to CEN/CENELEC for the provision, as preferred means of compliance, of voluntary, harmonised standards that will not fall foul of safeguard clauses or the general application of article 36 of the Treaty.

2.8.4 *CE* marking

The New Approach Directives provide for a single, distinctive marking (the letters "***CE***") to be displayed on products marketed as conforming to the applicable essential requirements, irrespective of the particular standards and conformity assessment procedures applied. Confidence in ***CE*** marking depends on the effectiveness of its enforcement across Europe.

Building materials may be tested in any Member State by a recognised testing laboratory and will carry the ***CE*** mark. Their use will be permitted without any further test requirements.

2.9 The Construction Products Directive (89/106/EEC)

The Council Directive of 21 December 1988 on the approximation of laws, regulations and administrative provision of the Member States relating to construction products (89/106/EEC).

One of the aims of the Directive is the removal of technical barriers to trade in the construction field, in so far as they cannot be removed by means of mutual recognition among Member States.

Within each Member State there are Regulations or other forms of building control that represent barriers to the free movement of goods and services throughout the European Union. The requirement to undertake different test procedures in each Member State in order to place construction materials on the market places a considerable burden on manufacturers in terms of actual costs incurred in getting their products certified and appropriate in terms of performance to gain local acceptance.

In brief, the objective of this Directive is to provide a series of harmonised testing standards, by which construction materials can be tested and classified. Once the classification criteria have been met, the manufacturer may affix the CE mark to his product. Under the provisions of the Directive, that product may then be placed on the market in all the Member States without the need for further testing.

All construction work within the European Union will eventually be controlled using a single set of harmonised standards (hENs). British Standard BS476 will disappear, as we know it today.

2.9.1 The Essential Requirements

On 7 May 1985, the council of Ministers agreed on a "New Approach To Technical Harmonisation and Standards", to tackle this long-standing problem in a new, quicker and more effective way. "New approach" directives set out "essential requirements" written in general terms, which must be met before products can be sold in the United Kingdom or anywhere else in the community. European standards' bodies fill in the detail and will be the main – but not the only – way for business to meet the essential requirements. Products meeting the requirements are to carry the ***CE*** mark. Therefore, products meeting the requirements may be sold legally in the United Kingdom or anywhere else in the Community. It is accepted that harmonised standards would lead to mutual acceptance of other country certification.

It is those product directives which state one of their essential requirements as being fire, flammability etc., which are of relevance to the fire services and such directives form an important part of teaching and researching fire precautions.

2.9.2 The Interpretative Documents (Explanation of the Essential Requirement 'Safety in Case of Fire')

For the purposes of the Construction Products Directive, the following two important definitions apply:

(a) Construction Works

Construction works is everything that is constructed or results from construction operations and is fixed to the ground. This term covers both buildings and engineering works. It refers to the complete construction comprising both structural and non-structural elements.

(b) Construction Products

This term refers to products, which are produced for incorporation in a permanent manner in the construction works. The term includes material, components, and elements, prefabricated systems and installations that enable the construction works to meet the essential requirements.

Many different reaction to fire tests (about 37) have been developed historically in the Member States and form the basis of various national regulations. The difference between national reaction to fire tests, used as a basis for the classification of products, is recognised as an obstacle to the internal trade of products. The tests do not lend themselves to harmonisation because they have evolved for prescriptive purposes for the particular needs of Member States without a firm scientific base or common measurement principles.

The reaction to fire hazard which the Construction Products Directive seeks to address is mainly that rising from the spread of fire and smoke within a construction works. This is complex because the hazard depends upon many factors including the nature and type of products (wall and ceiling linings, floor coverings, insulation etc.) and the use of the construction works, the type of room (small, large, corridor etc.) and the stage of the fire development. The reaction to fire of external components of the construction works such as roofs and facades is also important.

Fire safety requirements constitute a vital part of the regulations for works in the EU Countries. Fire safety in construction works includes requirements on the layout of buildings and on the performance of structures, building products, services and installations and fire safety installations under fire conditions.

Fire safety objectives are concerned (within reasonable limits) with safety of people in and around buildings, including the need to prevent fires. The requirements to satisfy these objectives automatically provide a degree of protection to the property.

An important part of the strategy is to minimise the occurrence of fire (fire prevention) but this document cannot cover all the relevant factors, such as, for example, fire safety management.

The development and growth of fire is complex. It depends upon a number of factors including the nature and distribution of the contents (fire load), the air supply, the thermal properties of the enclosure of the construction works, the fire and smoke control systems, and the fire protection system efficacy. Building contents however, are not a matter for this Directive. The reaction to fire performance of the internal linings of a room (its wall and ceiling surfaces and to some extent its floor coverings) can influence the rate at which fire and smoke develop and therefore is often controlled.

In addition fire safety of the occupants can be improved by early detection of a fire, which may be provided by an automatic fire detection and alarm system and/or by suppression of fire by appropriate fire protection system

As stated in Article 3 of the Directive, the purpose of the Interpretative Documents is to give concrete form to the essential requirements for the creation of the necessary links between the essential requirements and the mandates for the preparation of harmonised standards and guidelines for European technical approval.

2.10 Interpretative Document 2 – Safety in Case of Fire

This document addressed the essential requirement laid down in the Directive as follows:

'The construction works must be designed and built in such a way that in the event of an outbreak of fire:

(a) The loadbearing capacity of the construction can be assumed for a specific period of time;
(b) The generation and spread of fire and smoke within the works is limited;
(c) The spread of fire to neighbouring construction works is limited;
(d) Occupants can leave the works or be rescued by other means;
(e) the safety of rescue teams is taken into consideration.'

Fire safety engineering covers the way in which fire safety in a construction works is evaluated by means of calculation methods, taking into account the performance and effects of products, i.e. passive and active fire protection measures.

Fire safety engineering includes a number of activities of wide scope which influence the safety in use of construction works and construction products. It includes the use of calculation methods for determining the development and spread of fire and smoke, the time for untenable conditions to be reached and evacuation time. It also includes the determination of actions (thermal and mechanical) acting on products in case of fire, the determination of performance of passive and active fire protection measures and the design of these measures, e.g. fire safety installations and fire-resisting building elements.

In a fully integrated approach, fire safety engineering may include the complex interaction between the performance of passive and active fire protection measures and the construction works, i.e. development of life threatening atmospheres and the people behaviour to achieve safety in the most flexible and cost effective way. An engineered approach requires that relevant characteristics of products are provided, and calculations and design procedures are validated on an agreed and harmonised basis.

2.11 Mandated Standards

A Mandate is a mechanism by which the Commission secures the provision of harmonised technical standards (hENs) from the European Standards body. They are required to give substance to the Essential Requirements of the Directive. Once the European Standards body publishes harmonised test methods, the rules require that any conflicting National Standard must be withdrawn.

The rules however, allow a certain period of co-existence, but the Commission recommends that this be kept to an absolute minimum. The basic objective remains that all conflicting National Standards are withdrawn.

The objective therefore, of a Mandate is to require the European Standards body (CEN) to lay down provisions for the development of, and the quality of harmonised European standards in order, on the one hand, to make 'approximation' of national laws, regulations and administrative procedures possible, and on the other hand, to allow products conforming to them to be presumed to be fit for their intended use, as defined in the Directive.

2.12 The Standing Committee on Construction

Within the Provisions of the Construction Products Directive are requirements for the Commission to establish a Standing Committee to oversee the implementation of the essential requirements of that

Directive. Chaired by officials from the Commission (DG – Enterprise), it comprises representatives from the Member State Governments with responsibility for building control (in the United Kingdom it is serviced by officials from the Department of the Environment, Transport and the Regions).

The Standing Committee has decision-making powers in three areas:

(a) Classes of requirements and methods of attestation, to be specified in mandates;
(b) Interpretative documents; and
(c) Recognition of national technical specifications.

The European Commission proposes draft measures to the Committee. If the Committee concurs, the Commission goes ahead. If not, the Commission may submit a proposal to the Council of Ministers for a decision.

A sub-group known as The Regulators Group advises it.

2.12.1 The Regulators Group

Reporting directly to the Standing Committee, the Regulators Groups comprise a small number of representatives (one from each Member State). The membership comprises representatives of those Government departments in the Member States, who maintain the responsibility for the regulation of building control. In the United Kingdom, The Department of the Environment, Transport and the Regions undertakes this function. The function of this group is to deal with the detail of the 'fire' requirements of the Directive and to oversee the day-to-day work in connection with the development within CEN of the Mandated technical standards. The Laboratory Group assists it in this.

2.12.2 The Laboratory Group

Reporting to the Regulators Group, this small group comprises a single representative of those official laboratories within the Member States who are involved in the necessary research work to develop the Mandated harmonised technical standards. In the United Kingdom, the Fire Research Station undertakes this task.

2.13 Directive 98/34/EC (Formerly 83/189/EEC)

Arrangements have been in place since 1989, under the above Directives, to prevent the creation of new technical barriers to trade. Member States are required to notify the European Commission, in advance, of draft proposals for new technical regulations. This gives the Commission and other Member States an opportunity to intervene if they judge the proposed regulation would act as a barrier to trade.

A standstill of three months is then automatically imposed. This can be extended by three months if the Commission or another Member State argues that the proposals would create a new barrier to trade – a 'detailed opinion' – or to a year if the Community decides to propose Community-wide measures.

The European Community and the European Free Trade Association (EFTA) countries have agreed to exchange notifications and to provide for making comments on each other's notifications during the initial three-month standstill period, but not for any extension of that period.

2.14 Standard Terms and Definitions

In any discussion on fire safety, certain terms and definitions are used. This terminology will be used by the many agencies that promote fire safety in buildings and it is important therefore to have a thorough grasp of all such language. This is particularly important when dealing with technical standards, as definitions ascribed can be crucial in appreciating the many parameters of the Standards themselves.

In order to achieve the common use and acceptance of a standard terminology, the European and International standards bodies have produced a document (BS EN ISO 13943 – Fire Safety – Vocabulary) giving the internationally agreed terminology. This document is accepted by the United Kingdom and has been accorded a BS Number. Whilst the current document contains a large number of agreed definitions, it is a dynamic document in that it is not yet complete. The long-term objective is to produce a Dictionary of all terms and definitions related to 'fire'.

Where appropriate to the text of this publication, it is the Standard definitions that are used. However, in addition to the above document, the 2000 Edition of The Building Regulations 1991 – Fire Safety – Approved Document B contains a list of Definitions in Appendix 'E'.

Some of these definitions may differ from those in BS EN ISO 13943; where such differences are identified, the definitions given in Approved Document 'B' should be used until they are incorporated in the Standards document.

Chapter 3 – Basic Building Construction

It is necessary for firefighters to have some basic knowledge of the principal methods of building construction.

This chapter deals with some of the constructional methods used and gives examples of both traditional and newer types of buildings (Figure 3.1).

3.1 Constructional methods

The possibility of collapse of a building at an incident has been an ever-present problem for firefighters and an ability to assess a situation is important. Any large fire or explosion can make a building potentially dangerous as what remains will be under a great deal more stress than usual.

That is why Safety Officers are detailed at an incident for the specific purpose of detecting dangerous conditions and warning the Incident Commander in time for him to make a decision on what action to take to safeguard personnel. Some of the traditional signs of stress are discussed in other parts of this publication, but modern buildings may include lightweight roofs, cladding (both heavy and light), curtain walling, large areas of glass or polycarbonates in relatively light framing, cantilevered support structures etc.

Their behaviour has only been fire-tested usually as individual elements of structure; the reaction of the whole building, together with its internal fire loading, has not.

Building Regulations usually attempt to redress this situation by requiring the building to be stable and remain stable under fire conditions. This is achieved by dividing the building into fire components and specifying a degree of fire resistance for the elements of structure forming that compartment appropriate to the size, height and use of that compartment.

3.1.1 Solid construction

Solid construction, often referred to as "traditional" or masonry construction, consists of loadbearing external walls which support the floors and roof. The materials most commonly used were brick, concrete blocks or stone. This form of construction was almost universal during, and before, the 19thC for all kinds of industrial and commercial buildings. The example shown in Figure 3.1 is a warehouse but factories, cotton and woollen mills and old office blocks are all similarly built.

The walls are of solid brick or stone possibly up to one metre thick or more at the base but setting back on the upper floors. Cast-iron columns support either cast-iron or timber beams which are bedded into the loadbearing walls at either end. Where there are timber beams the commonest type of floor is made of 50 to 70mm solid wooden boards spanning from beam to beam.

Many older buildings (in which cast-iron beams may have been used) have barrel vaults and a few of the later ones have concrete floors on filler joists. The commonest roof is slated and pitched on wooden rafters and purlins supported by roof trusses.

3.1.2 Structural steel frame construction

This type of building has a great advantage from an architect's/designer's point of view in that, in its construction, the load of the floors and cladding is carried at each level by beams which, in turn, pass the load on to the columns.

Within a skeleton framework floor space, divided in a variety of ways, can be provided and a suitable non-loadbearing cladding material used as a weather and insulation wall. The skeleton of the building (Figure 3.3) is made up of universal

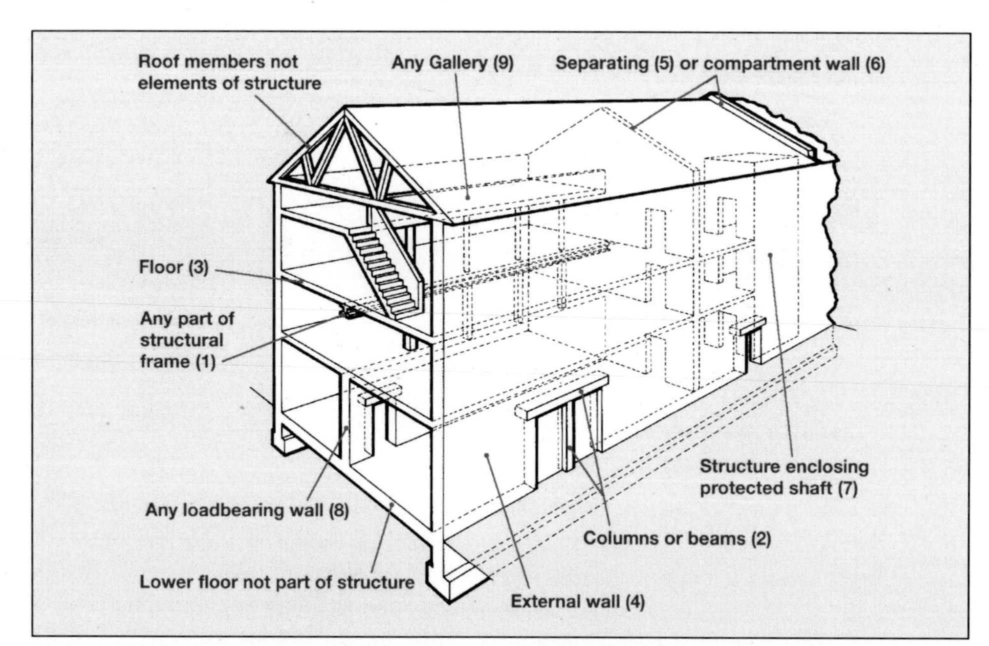

Figure 3.1 Sketch showing the various elements of structure of a building.

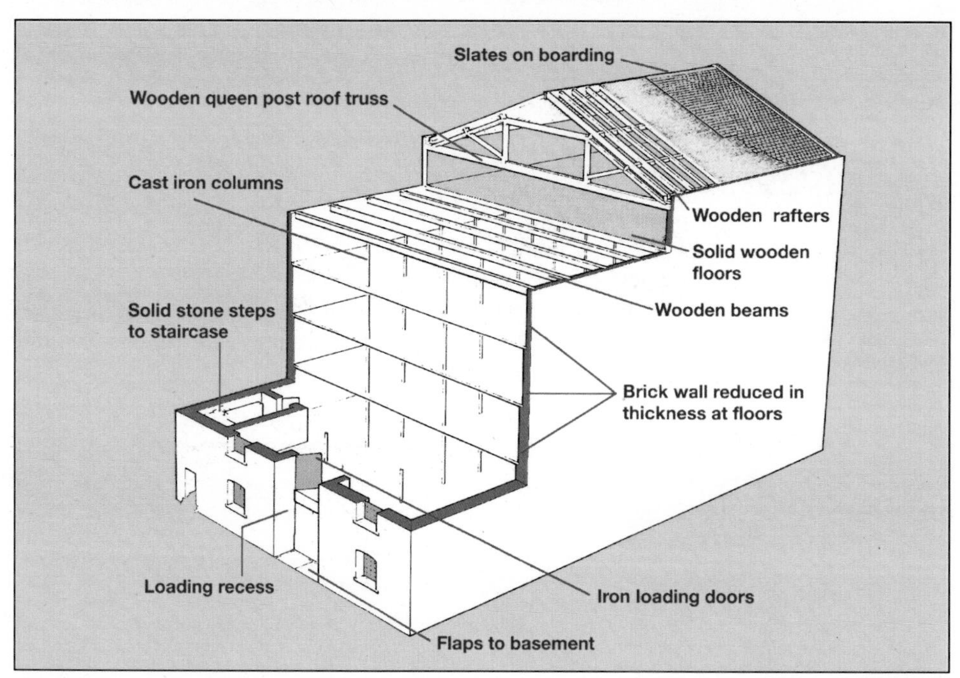

Figure 3.2 Typical form of solid construction with load-bearing walls used during the 19th century for largebuildings such as mills, warehouses, etc.

beams and columns designed to support the loads (static and rolling) on the floors, the cladding and the wind pressure. How the columns are arranged is usually determined by the various circulation spaces in the building and, to some extent, the window openings (see Figure 3.4).

The steel work would, normally, be required to be protected against fire by either "solid" or "hollow" protection.

The technical advances in steel and its use in construction has enabled designers to be very innovative. It becomes difficult to equate a steel "frame" with some of the buildings already erected and in the process of erection. Huge frameworks are often assembled using tubular steel in a highly complicated manner and the remainder of the building is "hung" onto it. These examples may range from large multi-storey office blocks to some acres of one and two storey shopping and leisure centres.

An example is given in Figure 3.5 of a building virtually cantilevered out on cables from a tower at one end and, in Figure 3.6 is a similar, slightly more complicated, version.

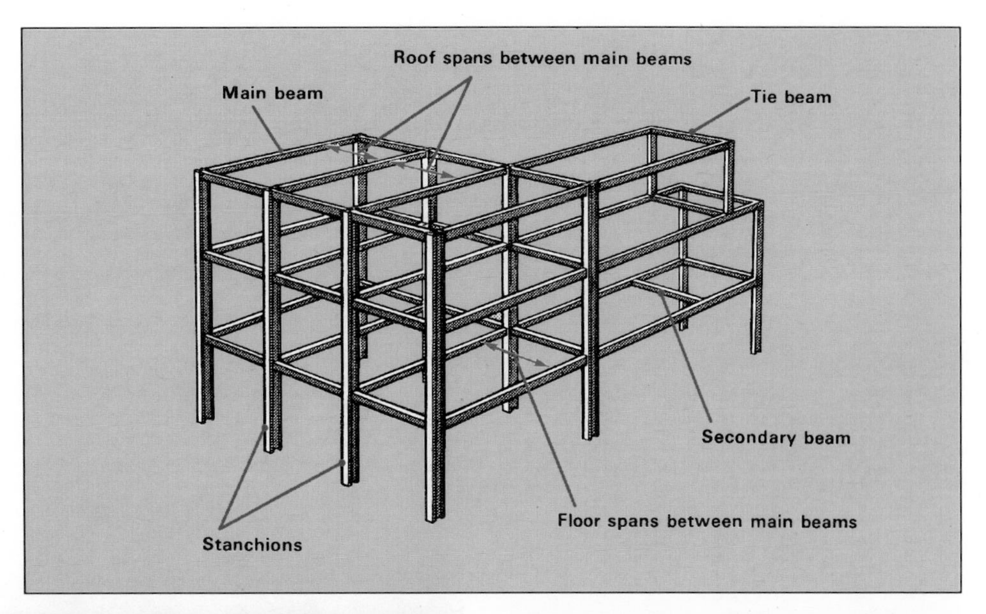

Figure 3.3 An example of a structural steel frame for a building.

Figure 3.4 Steelwork of various sizes in a framed building. Part concrete "plank" and part poured concrete floor. Note concrete cladding on lower column. Photo; Essex Fire Brigade.

Figure 3.5 A large roof basically supported by cantilever from a steel framed tower. Photo: Nicholas Grimshaw and Partners Ltd.

These combinations of "traditional" steelworks, tubular steelworks and "light" steel, or other metal, frames, added to these frames, large areas of glass or polycarbonates, metal-brick-concrete-plastic cladding, a variety of roofs including stretched fabrics (see Figure 3.6), plus the internal fire loading and it becomes obvious that fire-fighters need to look at buildings with more than a passing interest.

3.1.3 Reinforced concrete construction.

The reinforced concrete frame was, when first used, treated as an alternative to steel frames, i.e. the columns supported the main beams which, in turn, supported the floor slabs. This, however, gradually changed to a monolithic type of construction (Figure 3.7) where the columns, beams and floors were cast integrally.

There is a trend back to the original concept where the concrete floors are concrete slabs or planks laid between the beams (see Figure 3.8) . Another variation is to lay metal shuttering between beams and then, leaving the shuttering in place, lay a concrete floor on top.

Figure 3.6 A supermarket showing the method of suspended roof support. Photo: Ernest Ireland Construction.

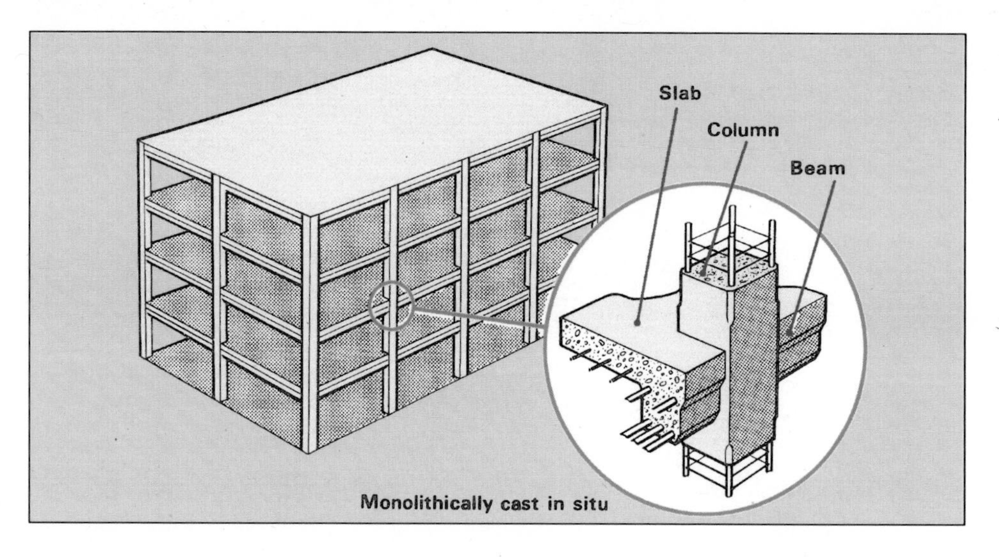

Figure 3.7 In situ reinforced concrete frame construction.

3.1.4 Precast reinforced concrete frame

Here the reinforced concrete frame components are manufactured at the factory and then assembled on site in a similar manner to steel frames (see Figure 3.9).

3.1.5 Composite construction

In this case the technology of lightweight structural steelworks is combined with the strength of precast concrete columns. Figure 3.10 shows an example.

3.1.6 Modular systems

The differences between modular systems, composite construction and precast construction are blurred but the main advantage of most modular systems is that, within certain parameters, prefabricated components can be used in an almost unlimited variety of ways. This includes their positioning for varying floor heights, spans, vertical spacing etc. plus an ability to use the same cross-sectional component in different loading conditions. Most

Figure 3.8 Type of concrete "plank" showing its final position right. Photo: Essex Fire Brigade.

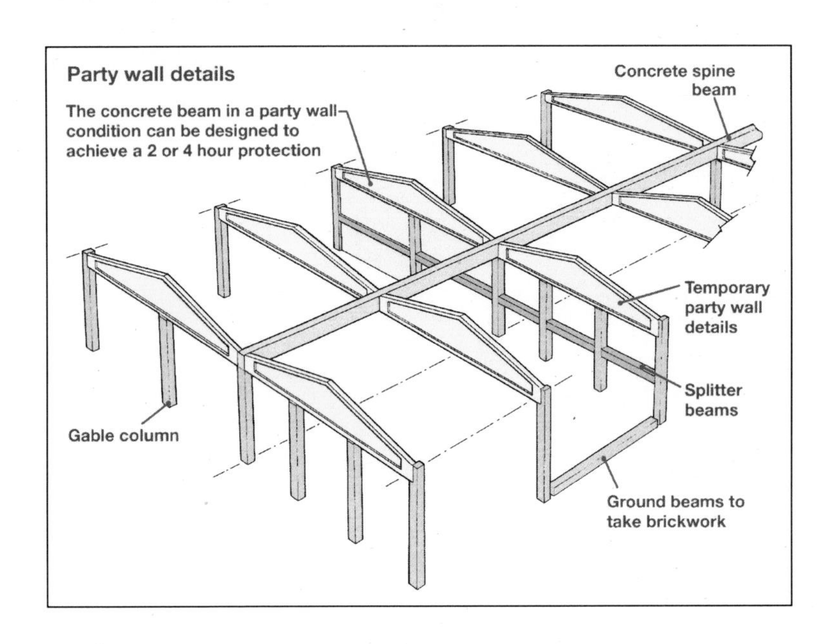

Figure 3.9 Typical reinforced concrete frame building.

systems use specially designed connectors with which to assemble the building.

These can be moulded into the precast columns or beams at whatever position suits the design of the building. An example of a connector system by Trent Concrete is shown in Figure 3.11. The steelworks in these components is usually encased in concrete and the steel connectors are covered in concrete in situ giving a degree of protection against fire and corrosion.

Figure 3.12 illustrates a multi-storey building, totally modular including floors, cladding, spandrels, stairs, columns and beams.

Figure 3.13 illustrates another type of modular building which includes modular roof sections.

3.1.7 Lift-slab construction

In this system the columns are constructed, the roof slab is formed and hydraulic jacks lift it first to the top of the column. Other floor slabs are formed and these too are lifted and "parked" on the columns. The sequence of lifting cycles is repeated until the structure is complete.

The columns are extended by "splicing" on another as the need arises. Figures 3.14–3.16 illustrate such a building in some of its stages of construction.

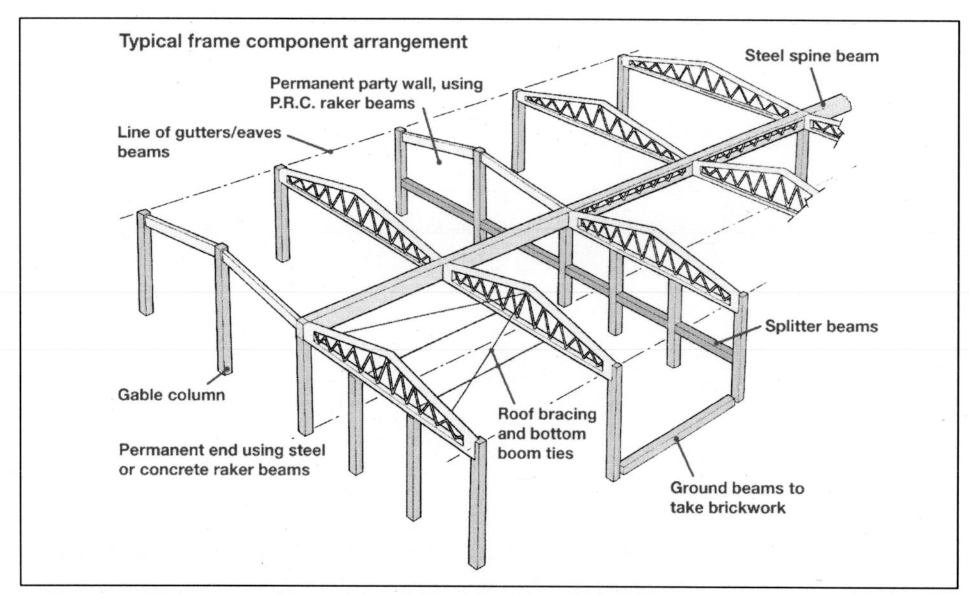

Figure 3.10 Combination of light-weight steelwork and precast concrete columns.

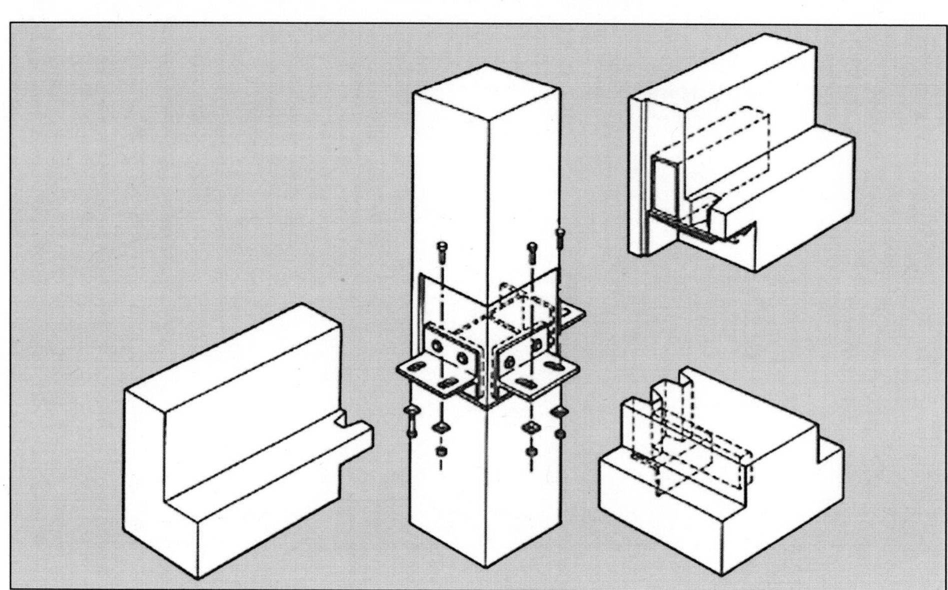

Figure 3.11 Typical connector system for modular building.

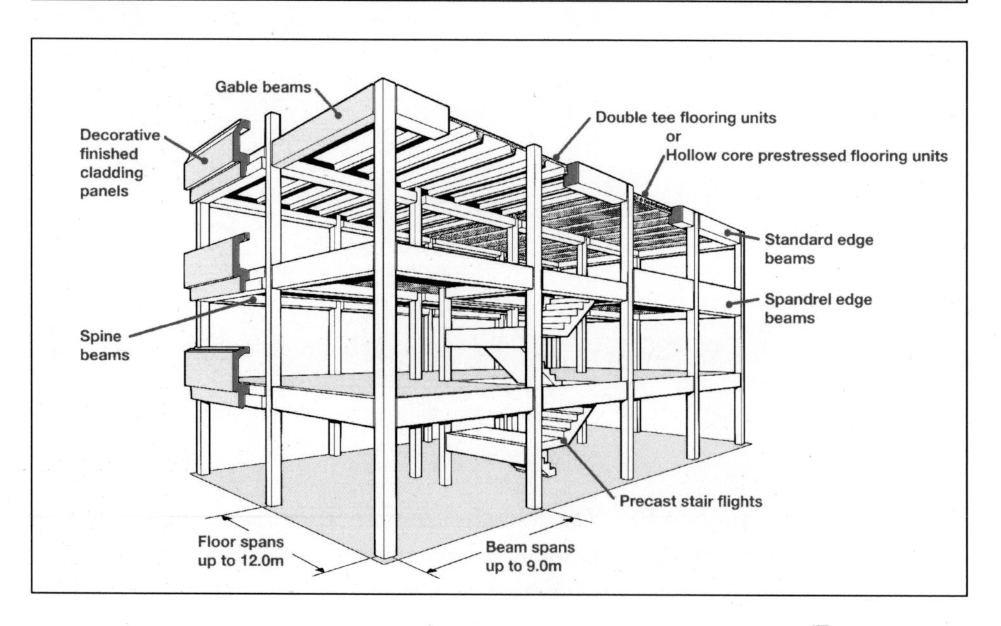

Figure 3.12 A totally modular building.

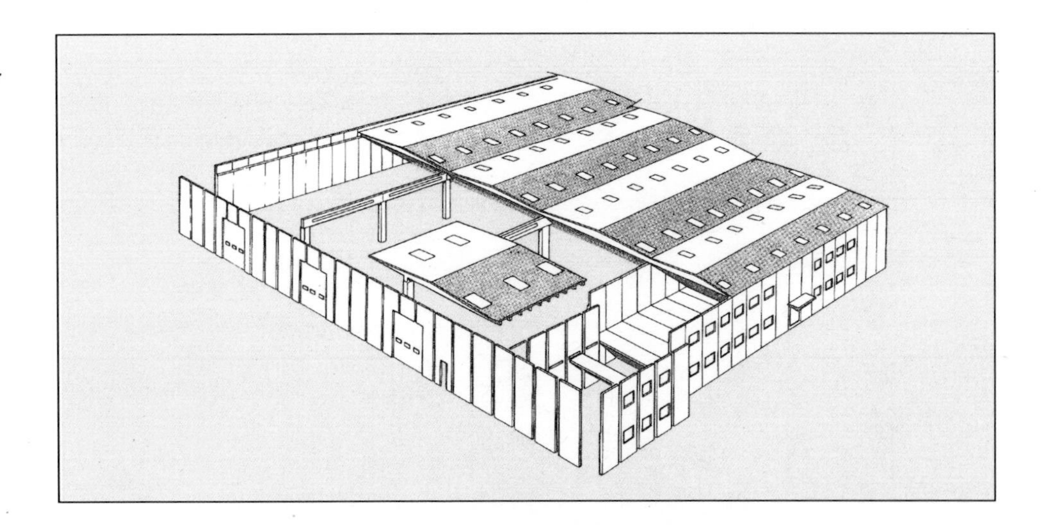

Figure 3.13 Another example of a modular building system.

Figure 3.14 Early stage in a lift-slab construction.
Photo: Douglas Specialist Contractors Ltd.

Figure 3.15 Lift-slab construction in its middle stages.
Photo: Douglas Specialist Contractors Ltd.

Figure 3.16 Lift-slab construction with all floors in position and shell well advanced.
Photo: Douglas Specialist Contractors Ltd.

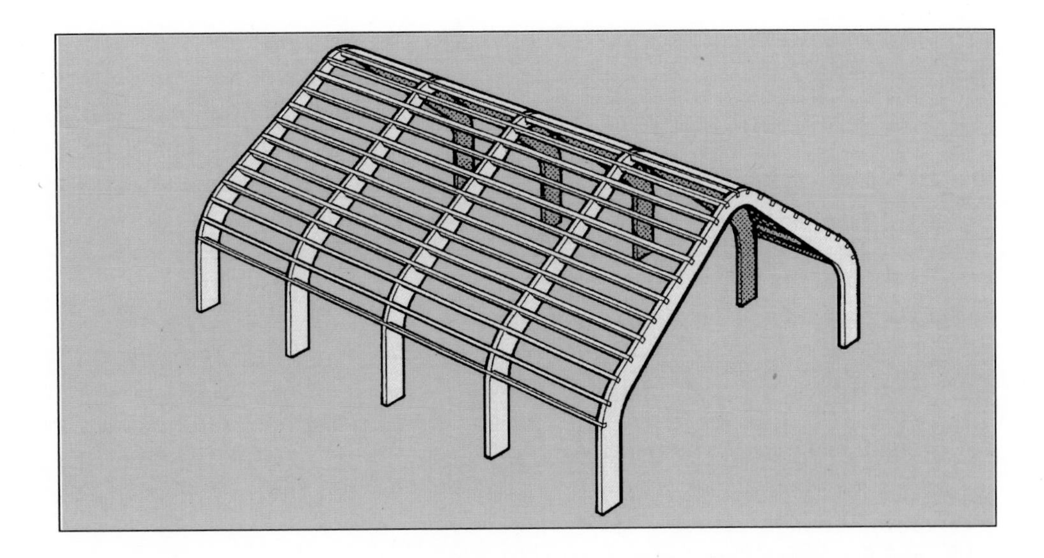

Figure 3.17 Diagram of a laminated timber Portal frame for a building.

3.1.8 Portal frame construction

This type of construction has largely been superseded by composite or modular construction but is still utilised satisfactorily using either concrete, steel or glulam timber methods. The columns and roof members are continuous requiring little or no internal bracing and supporting the roof on a series of purlins (see Figure 3.17). This gives a large unencumbered storage or working area.

3.1.9 Insulating Sandwich Panels

Sandwich panels have been implicated in a number of fires in recent years and concerns have been expressed regarding the risks that they present both to occupants and firefighters.

Fire Behaviour of Insulating Core Panels used for Internal Structures (Sandwich Panels)

Introduction

The following addresses the provision of guidance for fire prevention officers when they are consulted by building control bodies, occupiers or their agents, with regard to premises in which it is proposed to install, or which currently contain, internal structures constructed with insulating core panels.

Background

The most common use of insulating core panels, when used for internal structures, is to provide an enclosure in which a chilled or sub-zero environment can be generated for the production, preservation, storage and distribution of perishable foodstuffs. However, this type of construction is also used in many other applications, particularly where the maintenance of a hygienic environment is essential.

These panels typically consist of an inner core sandwiched between, and bonded to, a membrane such as facing sheets of galvanized steel, often with a PVC facing for hygiene purposes. The panels are then formed into a structure by jointing systems, usually designed to provide an insulating and hygienic performance. The panel structure can be free standing, but is usually attached to the building structure by lightweight fixings and hangers.

The most common forms of insulation in present use are:

(a) Expanded Polystyrene (Eps),
(b) Extruded Polystyrene (Xps),
(c) Polyurethane (Pu),
(d) Polyisocyanurate,(pIr),
(e) Mineral Fibre (MRf).

However panels with the following core materials are also in use:

(f) Modified Phenolic (Ph),
(g) Foam Glass (Fg).

(The initials contained in the brackets are those used in the Labelling and Certification Scheme referred to later in this section.)

Fire Behaviour of Core Materials and Fixing Systems

The behaviour of these panel systems, when involved in fire, differs in important respects from other construction systems. The nature of this behaviour and the implications for designers and approving bodies must be taken into account at the earliest opportunity.

Core Materials

Irrespective of the type of core material, the majority of panels (unless provided with appropriate mechanical fixings between the facings) will, when exposed to the high temperatures of a developed fire, tend to delaminate between the facings and core material, due to a combination of expansion of the membrane and softening of the bondline.

In addition the degradation of polymeric materials can be expected when subjected to radiated/conducted heat from a fire. This is likely to result in fire-spread within the panel and in the production of large quantities of smoke, before delamination has occurred. The delamination of polymeric-cored panels can also add to rapid fire-spread and lead to flashover conditions.

Fixing Systems

Once it is involved, either directly or indirectly in a fire, it can be anticipated that the panel will have lost most of its structural integrity. The stability of the system will then depend on the residual structural strength of the non-exposed facing, the joint between panels and the fixing system. If the ceilings of these systems are being used to carry loads, such as refrigeration or air-conditioning plant, the hazard caused by the collapse of the panel system will be further exacerbated.

Most jointing or fixing elements for these systems have an extremely limited structural integrity performance in fire conditions. If the fire starts to heat up the support fixings or structure to which they are attached, then there is a real chance of total collapse of the panel system.

General

The nature of these panel systems means that fire can spread behind the panels, hidden from view. The panels and fixing systems are, therefore, susceptible to the effect of fire from a number of directions, by means of conduction, convection and radiated heat.

Whilst it is recognised that the potential for problems in fires involving mineral fibre cores is less than those for the polymeric materials, the potential hazards caused by the collapse of the system, and hidden fire-spread, are common to all panels irrespective of the type of core.

Problems For Firefighting

Therefore, as already highlighted above, when compared with other types of construction techniques these panel systems provide a unique combination of problems for firefighters, including:

(a) Hidden fire-spread within the panels,
(b) Production of large quantities of black, toxic smoke and
(c) Rapid fire-spread leading to flashover.

These three characteristics are common to both polyurethane and polystyrene cored panels, although the rate of fire-spread in polyurethane cores is significantly less than that of polystyrene cores, especially when any external heat source is removed.

In addition, irrespective of the type of panel core, all systems are susceptible to:

(a) Delamination of the steel facing,
(b) Collapse of the system,
(c) Hidden fire-spread behind the panel system.

3.1.10 Design of Buildings Containing Insulating Core Panels

Existing Panels

Implementing the available guidance together with the development of the labelling scheme, will undoubtedly contribute to improving the fire safety of both occupants and firefighters should a fire occur in a building constructed with internal sandwich panels systems. However, it must be recognised that it will take some time for these benefits to be realised and that, in the interim, the majority of panel systems will not meet the performance requirements now being suggested.

Officers in charge of fire-ground operations must, therefore, continue to be mindful of the hazards presented by this type of construction.

3.1.11 Design Recommendations for Buildings Containing Insulating Core Panel Systems

It must be borne in mind that panels are not themselves the cause of fire. There are neither good nor bad panel cores, only badly selected cores or poorly detailed construction. Each design or application should be considered on its merit taking into account the complete design needs for the element of structure and the characteristics of the various panel cores.

To identify the appropriate solution a risk assessment approach should be adopted. This would involve identifying the potential fire risk within the enclosures formed by the panel systems, and then adopting one or more of the following at design stage:

(a) Removing the risk,
(b) Separating the risk from the panels by an appropriate distance,
(c) Providing a fire-suppression system for the risk,
(d) Providing a fire-suppression system for the enclosure,
(e) Providing panels with non-combustible cores,
(f) Specifying appropriate materials/fixing and jointing systems.

In summary, the performance of the building structure, including the insulating envelope, the superstructure, the substructure, etc, must be considered together in relation to their performance in the event of a fire.

Specifying panel core materials

Where possible, the specification of panels with core materials appropriate to the application will help ensure an acceptable level of performance for panel systems, when involved in fire conditions.

The following are examples of core materials that may be appropriate to the application concerned.

Non-Combustible Cores

(a) Cooking areas,
(b) Hot areas,
(c) Bakeries,
(d) Fire-breaks in combustible panels,
(e) Fire-stop panels.

Core materials may be used in circumstances other than those outlined above, where a risk assessment has been made and other appropriate fire precautions have been put in place.

General fire protection

All other cores

(a) Chill stores;
(b) Clean rooms;
(c) Cold stores; *
(d) Blast freezers; *
(e) Food processing factories. *

For those applications indicated by an asterisk (*), food safety and hygiene are the principal factors that will determine the selection of core materials.

3.1.12 Specifying Materials/Fixing and Jointing Systems

The following are methods by which the stability of panel systems may be improved, in the event of a fire, although they may not all be appropriate in every case.

(a) The details of construction of the insulating envelope should, particularly in relation to combustible insulant-cores, prevent the core materials from becoming exposed to the fire and contributing to the fire-load.

(b) Insulating envelopes, support systems, and supporting structures should be designed to allow the envelope to remain structurally stable by alternative means, such as catenary action, following failure of the bond-line between insulant core and facing material. This will typically require positive attachment of the lower faces of the panels to supports.
(c) The building superstructure, together with any elements providing support to the insulating envelope, should be protected to prevent early collapse of the structure or the envelope.
(d) Fixing systems using low-melt point elements, such as aluminium, should not be used.
(e) In designated high-risk areas consideration should be given to incorporating non-combustible insulant-cored panels into wall and ceiling construction at intervals, or incorporating strips of non-combustible material into specified wall and ceiling panels, in order to provide a barrier to fire propagation through the insulant.
(f) Correct detailing of the insulating envelope should ensure that the combustible insulant is fully encapsulated by non-combustible facing materials that remain in place during a fire.
(g) The panels should incorporate pre-finished and sealed areas for penetration of services.

Irrespective of the types of panel provided, it will remain necessary to ensure that the supplementary support method supporting the panels remains stable for an appropriate period under fire conditions.

It is not practical to fire-protect light gauge steel members, such as purlins and sheeting rails, which provide stability to building superstructures, and these may be compromised at an early stage of a fire. Supplementary fire-protected heavier gauge steelwork members could be provided at wider intervals than purlins to provide restraint in the event of a fire.

3.1.13 General

In addition to the above the following general points should be borne in mind.

(i) Panels or panel systems should not be used to support machinery or other permanent loads.
(ii) Any cavity created by the arrangement of panels, their supporting structure or other building elements, should be provided with suitable cavity barriers.

Chapter 4 – Reaction to Fire Tests and Fire Resistance Tests

Introduction

These two parameters form the basis of the control of materials used in the construction of buildings, be they domestic dwellings or the most complex structures. Most of the tests are developed to measure one or more of the individual aspects such as ignitability, heat release, surface spread of flame; others have been developed to measure the rate of production of smoke and toxic gases.

Over the years, a great number of such tests have been developed based upon quite different characteristics. The heat insult may be supplied by an electrical heater, a gas flame or a flame from an oil burner. The rate at which the heat is applied to the specimen can be different as well as the final measurement and characterisation of the performance. Based upon such widely differing measurements, it is obvious that there can be little or no correlation between the results obtained by the use of such tests.

The need for a single, harmonised test protocol has long been recognised and the achievement of such harmonisation within the single European market has long been a priority of the European Commission.

The construction material producers have also been concerned over the burden placed upon them in order to get their products accepted in the Member States. Each Member State has traditionally produced its own destinctinve test regime and manufacturers were faced with the cost of having their products tested and classified in each member state.

The Construction Products Directive has addressed this and the European Standards body has now produced the first of many hundred of technical standards needed to give substance to the Essential Requirements of the Directive.

The range of agreed technical standards covering the reaction to fire and fire resistance testing of construction materials is provided as an Annex at the end of this Chapter.

The use of plastics materials has long been commonplace in the construction industry. In fact this material alone has provided architects and designers with the facility to introduce design features that would be unachievable without modern plastics materials.

Today, the use of plastics is so large that a whole series of tests specifially designed to measure the fire properties of such materials has been developed

Whatever the material and whatever the form the testing takes, certain basic principles are usually measured. They may be summarised as follows:

4.1 Ignitability

If a material does not ignite, there is no fire. Therefore, low ignitability is the first defence in a fire. In fact, however, all organic materials do ignite, but the higher the temperature a material has to reach before it ignites, the safer it is, Thus it is possible to determine ignition temperatures using standard methods of test. There are two ways this can be conducted:

(a) Determine the time to ignition; or
(b) Measure the minimum heat input needed to ignite the material.

4.2 Ease of Extinction

Once ignited, the easier a material is to extinguish, the lower the fire hazard associated with it.

One of the most widely used small-scale tests is the limited oxygen index test, British Standard BS

EN ISO 4589. It gives the limiting concentration of oxygen in the atmosphere necessary for sustaining combustion (higher numbers reflect greater ease of extinction) This test is widely used for specifications, although its applicability to real-scale fires has been severely criticised.

4.3 Flame Spread

The tendency of a material to spread flame can be measured with a variety of tests. This is one of the most important parameters measured in the whole range of reaction to fire tests.

4.4 Heat Release

The key question in a fire is 'how big is the fire'. The one fire property that answers that question is the rate of heat release. A burning object will spread fire to nearby materials only if it gives off enough heat to ignite them. Moreover, the heat has to be released fast enough not to be dissipated or lost whist travelling through the cold air surrounding anything not on fire. Therefore fire hazard is dominated by the rate of heat release.

In fact, rate of heat release has been shown to be much more important than either ease of ignition, flame spread or smoke toxicity in controlling the time available for potential victims of a fire to escape. Fire victims often die from inhaling smoke and toxic gases, however, unless the fire becomes very big, there is unlikely to be enough toxic smoke to kill. Therefore, fire fatalities occur when the rate of heat release of the fire is sufficiently large to cause many (or even most) products in the room of origin of fire to burn.

4.5 Smoke Obscuration

Decreased visibility is a serious concern in a fire, because both escape from the fire and rescue by firefighters is more difficult. The main way in which a fire decreases visibility is by the release of smoke.

Decreased visibility is the result of a combination of two factors:

(a) How much material is burned in the fire; and
(b) How much smoke is released per unit of material burned.

Several empirical parameters have been proposed to compensate for incomplete sample consumption under testing conditions. One of them – known as the smoke factor, has been used with small-scale rate of heat release calorimeters. It combines the two parameters mentioned above (a & b).

4.6 Tests for Toxic Potency

Fire hazard is also associated with the toxicity of the smoke itself. This is usually measured in terms of the narcotic gases (e.g. carbon monoxide) and the irritant gases (e.g. hydrogen chloride) released. The most important toxic product in any fire is carbon monoxide which is produced by all organic materials when they burn, however, tests have shown that a 'cocktail' of nearly a hundred gases can be detected with the use of specialised equipment. The significance of the interaction between these materials is still being examined, however, it is sufficient to concentrate on the more common fire gases, e.g. carbon dioxide, carbon monoxide, hydrogen chloride, hydrogen cyanide, low oxygen and acrolein.

Chemical tests exist for determining the presence of such gases (or lack in the case of low oxygen) but such tests cannot predict the results of possible synergism or additive effects of such gases. This can only be achieved by the use of bio-assay tests (use of an animal model). Such tests are not routinely conducted for this purpose in the United Kingdom or within the European Union.

There have been isolated instances in the past where it has been suspected that certain materials have demonstrated an unusually high toxic potency. In such conditions limited studies have been completed to evaluate this and this has necessitated the use of a bio-assay model as pure chemical analysis was not able to provide data.

As mentioned above, these were isolated instances and given the current attitude to the use of bio-assay models it is unlikely they will ever be repeated.

4.7 Fire Resistance Testing

The ability of structural elements to continue to function when subjected to the effects of heat is defined as its fire resistance and this is normally measured in terms of time. It is the fire resistance of the assemblies, not just components, which must be evaluated.

The fire resistance of a component, or assembly of components, is measured by the ability to resist fire by retaining its loadbearing capacity, integrity and insulating properties. The loadbearing capacity of the assembly is its dimensional stability. The integrity of the assembly is its ability to resist thermal shock and cracking and to retain its adhesion and cohesion. The insulation offered by the material is related to its level of thermal conductivity. Fire resistance is normally defined under these three characteristics (loadbearing capacity, integrity and insulation) and given in minutes or hours of resistance.

In the case of elements of structure, then only stability and integrity are immediately essential. However, if the elements of structure are also acting to subdivide the building either horizontally (floors) or vertically (walls) to contain the fire, then the insulation is also important.

When considering the fire resistance of a structural assembly, the designer must be aware that there can be significant differences between the performance of assemblies under test conditions and in reality. Obviously, test samples are of the highest quality and workmanship; these standards will have to be repeated on site if the same level of fire resistance is to be achieved. Test samples are also new, and it is important that the fire resistance of the assemblies in place is not jeopardised by the effects of mechanical damage, weathering or thermal movement.

ANNEX

4.8 Reaction to Fire Tests

DD 246:1999 – Recommendations for the use of the cone calorimeter.

This Draft for Development (DD) examines the limitations of the cone calorimeter as currently used for building products and recommends ways in which some of these may be overcome for other types of products for other application areas.

BS ISO TR 5658-1:1997 – Reaction to fire tests – Spread of flames – Part 1: Guidance on flame spread.

This Technical Report provides guidance on flame spread tests for construction products. It describes the principles of flame spread and classifies different flame spread mechanisms.

BS ISO 9239-1:1997 – Reaction to fire tests – Horizontal surface spread of flame on floor-covering systems – Part 1: Flame spread using a radiant heat ignition source.

This British Standard reproduces verbatim ISO 9239-1:1997, including Technical Corrigendum 1:1997 and Technical Corrigendum 2:1998, and implements it as a national standard.

BS ISO/TR 11696-1:1999 – Use of reaction to fire test results – Part 1: Application of test results to predict fire performance of internal linings and other building products.

This Technical Report describes how information on basic values for ignition, spread of flame, rate of heat release and smoke can be used in fire growth models for internal linings and other building products to describe the fire hazard in a limited number of scenarios starting with fire development in a small room. Other scenarios include fire-spread in a large compartment and fire propagation down a corridor.

BS ISO/TR 11696-2 – Uses of reaction to fire test results – Part 2: Fire hazard assessment of construction products.

This part of ISO/TR 11696 provides guidance on the principles and use of fire test data and other relevant information concerning construction products and their end-use environment so that potential fire hazards and/or risks may be assessed. It suggests procedures for expressing results and how to interpret the data and to aid the fire hazard assessment process.

BS ISO TR 11925-1:1999 – Reaction to fire tests – Ignitability of building products subjected to direct impingement of flame – Part 1: Guidance on ignitability.

This Technical Report provides guidance on 'ignitability' tests for building products. It describes the principles of ignitability and characterizes different ignition sources.

The results of small-scale ignitability tests may be used as a component of a total hazard analysis of a specified fire scenario. It is therefore important that the flame or radiative source chosen is fully characterized so that relevant conclusion may be made from the test results.

Guidance given in this Technical Report may also have relevance to other application areas (e.g. building contents, plastics, etc.)

4.9 Tests for Fire Resistance

BS EN 1363-1:1999 – Fire Resistance Tests – Part 1: General requirements.

This part of EN 1363 establishes the general principles for determining the fire resistance of various elements of construction when subjected to standard fire exposure conditions. Alternative and additional procedures to meet special requirements are given in EN 1363-2.

The principle that has been embodied within all European Standards relating to fire resistance testing is that where aspects and procedures of testing are common to all specific test methods e.g. the temperature/time curve, then they are specified in this test method. Where the general principle is common to many specific test methods, but the detail varies according to the element being tested e.g. the measurement of unexposed face temperature, then the principle is given in this document, but the detail is given in the specific test method. Where certain aspects of testing are unique to a particular specific test method e.g. the air leakage test for fire dampers, then no details are included in this document.

BS EN 1363-2:1999 – Fire Resistance Tests – Part 2: Alternative and additional procedures.

This part of EN 1363 specifies alternative heating conditions and other procedures that may need to be adopted under special circumstances.

Details of the alternative hydrocarbon, slow heating and external fire exposure heating curves and the additional impact test and measurement of radiation procedures are included within this standard. Within the appropriate clause for each procedure is given an explanation as to why it may be necessary.

Unless one of the alternative heating regimes is specifically required, the standard temperature-time curve given in EN 1363-1 shall be used. Similarly, the impact test and measurement of radiation shall only be undertaken when they are specifically requested.

DD ENV 1363-3:2000 – Fire resistance tests Part 3: Verification of furnace performance.

This European Pre-standard describes a procedure for the verification of the thermal and pressure characteristics of fire resistance furnaces for the testing of separating element.

BS EN 1364-1: 1999 – Fire resistance tests for non-loadbearing elements – Part 1: Walls

This part of EN 1364 specifies a method for determining the fire resistance of non-loadbearing walls. It is applicable to internal non-loadbearing walls with and without glazing, non-loadbearing walls consisting almost entirely of glazing (glazed non-loadbearing walls) and other non-loadbearing internal and external non-loadbearing walls with and without glazing.

The fire resistance of external non-loadbearing walls can be determined under internal or external exposure conditions. In the latter case the external fire exposure curve given in EN 1363-2 is used.

BS EN 1364-2:1999 – Fire resistance tests for non-loadbearing elements – Part 2: Ceilings.

This part of EN 1364 specifies a method for determining the fire resistance for ceilings, which in themselves possess fire resistance independent of any building element above them. This Standard is used in conjunction with EN1363-1.

BS EN 1365-1:1999 – Fire resistance tests for loadbearing elements – Part 1: Walls.

This part of EN 1365 specifies a method of testing the fire resistance of loadbearing walls. It is applicable to both internal and external walls. The fire resistance of external walls can be determined under internal or external exposure conditions.

The fire resistance performance of loadbearing walls is normally evaluated without perforations such as glazing. If it can be demonstrated that the design of the opening is such that load is not transmitted to the perforation, then the perforation need not be tested in the loaded conditions.

BS EN 1365-2:2000 – Fire resistance tests for loadbearing elements – Part 2: Floors and roofs.

This part of EN 1365 specifies a method for determining the fire resistance of:

- floor constructions, without cavities or with unventilated cavities;
- roof constructions, with or without cavities (ventilated or unventilated);
- floor and roof constructions incorporating a glazed element;

with fire exposure from the underside.

This Standard is used in conjunction with EN 1363-1.

BS EN 1365-3:2000 – Fire resistance tests for loadbearing elements – Part 3: Beams.

This part of EN 1365 specifies a method for determining the fire resistance of beams with or without applied fire protection systems and with or without cavities. This Standard is used in conjunction with EN 1361-1.

Beams, which are part of a floor construction, are tested with the floor construction as described in EN 1365-2 and are subject to evaluation of integrity and insulation.

BS EN 1365-4:1999 – Fire resistance tests for loadbearing elements – Part 4: Columns

This part of EN1365 specifies a method for determining the fire resistance of columns when fully exposed to fire on all sides. This Standard is to be used in conjunction with EN1363-1.

BS EN 1366-1:1999 – Fire resistance tests for service installations – Part 1: Ducts.

This part of EN 1366 specifies a method for determining the fire resistance of vertical and horizontal ventilation ducts under Standardisation fire conditions. The test examines the behaviour of ducts exposed to fire from the outside (duct A) and fire inside the duct (duct B). This Standard is used in conjunction with EN 1363-1.

BS EN 1366-2:1999 – Fire resistance tests for service installations – Part 2: Fire dampers.

This part of EN 1366 specifies a method of test for determining the fire resistance of fire dampers installed in fire-separating elements designed to withstand heat and the passage of smoke and gases at high temperature. The Standard is used in conjunction with EN 1363-1.

BS EN 1634-1:2000 – Fire resistance tests for doors and shutter assemblies – Part 1: Fire doors and shutters.

This part of EN 1634 specifies a method for determining the fire resistance of door and shutter assemblies designed for installation within openings incorporated in vertical separating elements. Doors tested in accordance with this standard can be acceptable for certain lift landing door applications.

BS ISO 10294-2:1999 – Fire resistance tests – Fire dampers for air distribution systems – Part 2: Classification, criteria and field of application of test results.

This part of ISO 10294 provides the classification and appropriate criteria for the test procedure described in ISO 10294-1:1996, for the assessment of a fire damper to prevent the spread of fire and hot gases from one component to another. It also specifies the size of dampers to be tested.

BS ISO 10294-3:1999 – Fire resistance tests – Fire dampers for air distribution systems – Part 3: Guidance on the test method.

This part of ISO 10294 gives guidance on the application of the test method specified in ISO 10294-1:1996.

This test method is concerned with the assessment of a fire damper to prevent the spread of fire and hot gases from one compartment to another. It is not intended for dampers used only in smoke control systems.

It is applicable to fire dampers included in air distribution systems.

The test is not designed to test fire protection devices which only deal with air transfer application, or when a damper is used in suspended ceilings as the installation of the damper and duct may have an adverse effect on the performance of the suspended ceiling and other methods of evaluation may be required.

BS ISO/TR 12470:1998 – Fire resistance tests – Guidance on the application and extension of results.

Direct and extended applications of test results are the two possible ways to ensure that a modified element will have a good possibility of obtaining the same fire rating as that of the original tested specimen. In both cases these applications refer only to the fire rating that the building element can expect to reach if it were to be tested in a furnace according to the standard used for the reference test.

BS ISO 11925-2:1997 – Reaction to fire tests – Ignitability of building products subjected to direct impingement of flame – Part 2: Single flame source test.

This part of ISO 11925 specifies a test method for determining the ignitability by direct small flame impingement under zero impressed irradiance of building products used in a vertical orientation to assess the lowest level of performance.

BS ISO 11925-3:1997 – Reaction to fire tests – Ignitability of building products subjected to direct impingement of flame – Part 3: Multi-source test.

This part of 11925 deals only with a simple representation of a particular aspect of a potential fire situation typified by a flame playing directly onto a material, composite or assembly.

BS ISO/TR 14696:1999 – Reaction to fire tests – Determination of fire parameters of materials, products and assemblies using an intermediate scale heat release calorimeter (ICAL).

This Technical Report provides a method for measuring the response of materials, products and assemblies exposed in vertical orientation to controlled levels of radiant heating with an external igniter.

This test method is used to determine the ignitability, heat release rates, mass loss rates, and visible smoke development of materials, products and assemblies under well-ventilated conditions.

4.10 Fire Tests for Plastics

BS EN ISO 5659-1:1999 Plastics – Smoke generation – Part 1: Guidance on optical-density testing.

This guidance document constitutes part 1 of ISO 5659. Part 2 of this International Standard describes a static (or cumulative) single chamber test procedure. At present, the scope of this guide is limited to the test procedure described in part 2.

BS EN ISO 5659-2:1999 – Plastics – Smoke generation – Part 2: Determination of optical density by a single chamber test.

BS ISO/TR 5659-3:1999 – Plastics – Smoke generation – Part 3: Determination of optical density by a dynamic flow method.

This part of ISO 5659 specifies a method of measuring smoke production from the exposed surface of specimens of essentially flat materials, composites or assemblies not exceeding 25mm in thickness, when placed in a horizontal orientation and subjected to specified levels of thermal irradiance under forced ventilation conditions, with or without the application of a pilot flame. This method of test is applicable to plastics and may also be used for the valuation of other materials (e.g. rubbers, textile coverings, painted surfaces, wood and other building materials).

BS EN ISO 9773:1999 – Plastics – Determination of burning behaviour of thin flexible vertical specimens in contact with a small-flame ignition source.

This international Standard specifies a small-scale laboratory screening procedure for comparing the relative burning behaviour of vertically oriented thin and relatively flexible plastics specimens exposed to a low-energy-level flame ignition source. These specimens cannot be tested using method B of ISO 1210 since they distort or shrink away from the applied flame source without igniting.

BS EN ISO 10093:1999 – Plastics – Fire tests – Standard ignition sources.

This International Standard describes and classifies a range of laboratory ignition sources for use in fire tests on plastics and products consisting substantially of plastics. These sources vary in intensity and area of impingement. They may be used to simulate the initial thermal abuse to which plastics may be exposed in certain actual fire risk scenarios.

BS ISO 11907-1:1998 – Plastics – Smoke generation – Determination of the corrosivity of fire effluents – Part 1: Guidance

The present guidance document constitutes the first part of ISO 11907, the other parts of which describe one static and two dynamic test procedures. At the present time, the scope of this guide is limited to these three test procedures as indicated in the foreword.

BS ISO 11907-3:1998 Plastics – Smoke generation – Determination of the corrosivity of fire effluents – Part 3: Dynamic decomposition method using a travelling furnace.

This part of ISO 11907 specifies a test method for generating thermal decomposition products from plastic materials or products in an air stream and assessing the corrosive effects of these fire effluents on targets. It is not intended that the results be used to assess the corrosivity hazard of fire atmospheres.

BS ISO 11907-4:1998 Plastics – Smoke generation – Determination of the corrosivity of fire effluents – Part 4: Dynamic decomposition method using a conical radiant heater.

This part of ISO 11907 specifies a test method for measuring the corrosive effect, by loss of metal from a target, of the combustion effluents of plastic materials or products. The test method is intended for the evaluation of materials or products, for additional data to assist in the design of products, and development and research purposes.

Chapter 5 – The basic elements of structure

5.1 Beams

The beam is probably the oldest structural member. It is not hard to imagine primitive man dropping a tree across a stream to form the first bridge.

A beam transmits forces in a direction perpendicular to such forces to the reaction points (points of support). Consider a load placed on a floor beam. The beam receives the load, turns it laterally, divides it, and delivers to the reaction points.

The definition of a beam does not consider its attitude, that is, its vertical or horizontal orientation. Whilst beams are ordinarily thought of as horizontal members, this is not always the case. A vertical or diagonal member that performs the functions of a beam. Although it may have another name such as a rafter, is structurally a beam.

When a beam is loaded, it deflects or bends downward. The initial load is its own dead weight. The load placed upon it is a superimposed load. Some beams are built with a slight camber or upward rise so that when the design load is superimposed, the beam will be more nearly horizontal. Deflection (Figure 5.1) causes the top of the beam to shorten so that the top is in compression. The bottom of the beam elongates and thus is in tension. The line along which the length of the beam does not change is the neutral axis or plane. It is along this line that the material the beam is doing the least work.

5.1.1 Different terms are used to describe various beams:

A simple beam – is supported at two points near its ends. In simple beam construction (Figure 5.2), the load is delivered to the two reaction points and the rest of the structure renders no assistance in an overload situation.

A continuous beam – is supported at three or more points. Continuous construction (Figure 5.2) is structurally advantageous because if the span between two supports is overloaded, the rest of the beam assists in carrying the load.

A fixed beam – is supported at two points and is rigidly held in position at both points. This rigidity may cause collapse of a wall if the beam collapses and the rigid connection does not yield properly.

An overhanging beam – projects beyond its support, but not far enough to be a cantilever.

A joist – is a wood or steel or precast concrete beam directly supporting a floor.

A steel joist or bar joist – is a lightweight steel truss joist.

A girder – is a beam, of any material (not just steel) which supports other beams.

A built-up girder – is made of steel plates and angles riveted together, as distinguished from one rolled from one piece of steel.

A Spandrel girder – is a beam, which carries the load of the exterior of a framed building between the top of one window and the bottom of the window above.

A lintel – is a beam, which spans an opening in a masonry wall. Stone lacks tensile strength so it can only be used for short lintels.

A grillage – is a series of closely spaced beams designed to carry a particularly heavy load.

A cantilever beam – is supported at one end only, but it is rigidly held in position at that end. It

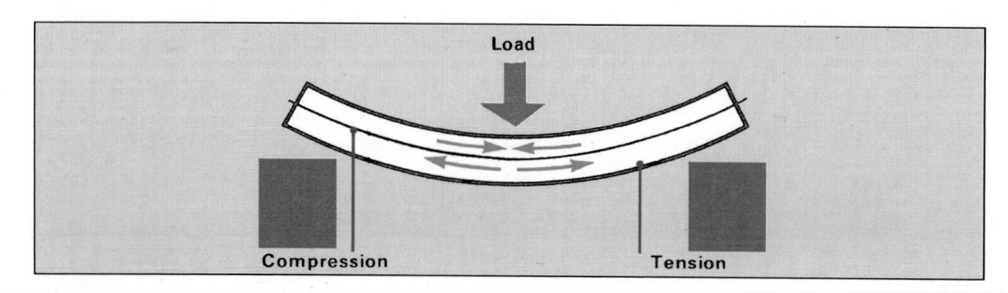

Figure 5.1 Diagram showing the effect of deflection on a beam. The curvature shown is greatly exaggerated.

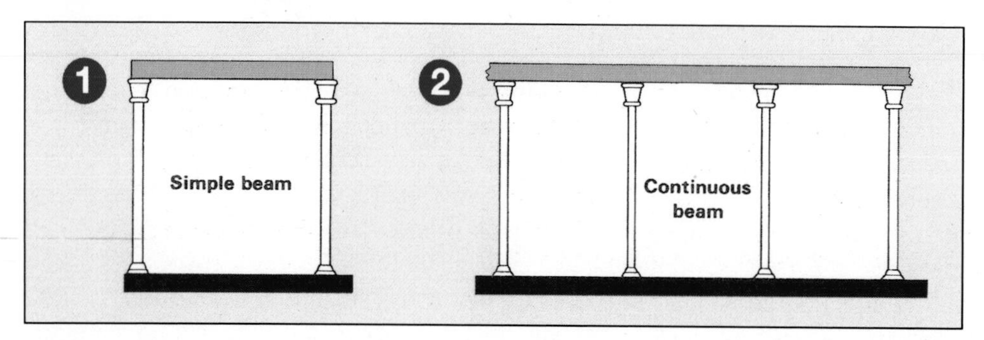

Figure 5.2 Diagram showing two types of beam: (1) a simple beam, (2) a continuous beam.

projects over a support point. Beyond the support point, the tension is in the top and the compression is in the bottom. Cantilever structures are very likely to be unstable in a serious fire because the fire may destroy the method by which the beam is held in place.

A needle beam – when any change is made in the foundations of an existing wall, the wall must be supported. Holes are cut through the wall, and needle beams are inserted and supported on both sides. They pick up the load of the walls.

A suspended beam – is a simple beam, with one or both ends suspended on a tension member such as a chain or rod.

A transfer beam – moves loads laterally when it is not convenient to arrange columns one above another. It changes the vertical alignment and is designed to receive the concentrated load of the column and deliver it laterally to supports.

5.1.2 Beam loading

Beam loading refers to the distribution of loads along a beam (Figure 5.2). Assume a given simple beam, which can carry eight units of distributed load. If the load is concentrated at the centre, it can carry only four units. If the beam is cantilevered and the load distributed, two units can be carried. If the load is at the unsupported end of the cantilever, only one unit can be carried.

5.1.3 Reaction and bending

The reaction of a beam is the result of force exerted by a beam on a support. The total of the reactions of all the supports of a beam must equal the weight of the beam and its load.

The bending moment of a beam can be simply described as that load which will bend or break the beam. The amount of bending moment depends not only on the weight of the load, but also on its position. The farther a load is removed from the point of support, the greater the moment; heavy loads should be placed directly over, or very close to the point of support.

5.1.4 Timber beams

Except in houses, solid timber is very seldom used in structural framing but could be found in the older industrial buildings. The charring rate of timber is generally accepted as 0.64mm per minute and an uncharred core is neither materially affected nor significantly reduced in strength. Recently there has been resurgence in the use of laminated (glulam) timber for public buildings. Spans of 150m are not uncommon using glulam made of European Whitewood.

The claimed charring rate for this glulam is 0.40mm per minute and the structural integrity is good because of the high fire resistance of the laminating adhesives. It is apparently possible to predict the inherent fire resistance of a component for a specified period and standards are set in the relevant British Standard. Class 0 surface spread of flame is usually obtained by the application of a proprietary treatment after the structure is erected.

5.1.5 Reinforced concrete

To compensate for its lack of tensile strength concrete high tensile steel rods reinforce beams. These are usually held in place by a designed system of steel latticework according to the type of construction strength required and then encased in concrete. Steel is used for reinforcement for three reasons:

(a) It can withstand high tensile stresses;
(b) The expansion rates of steel and concrete are almost the same; and
(c) The adhesion between the two surfaces in contact results in efficient bonding of the two materials.

These beams are inherently non-combustible but the fire resistance is dependent on the cross-sectional area of the beam and the amount of concrete cover provided for the reinforcement. Although reinforced concrete is a good structural material, being very strong and capable of almost limitless fabrication and flexibility in design, it has two main problems:

(i) The heavy deadload of the material results in limited effective spans of floors and beams unless specialist techniques are used.
(ii) Due to the low elasticity of concrete and the high elasticity of steel, soffits of floors and beams may crack on overload and the steel, if not properly protected, may be attacked by the capillary action of moisture etc.

This is one of the reasons why, in order to overcome these problems, a system of pre-stressing concrete was introduced.

5.1.6 Pre-stressed concrete

This method depends on the reliability of control of the aggregate and mix proportions, the placing of the concrete, the quality of the steel reinforcing and the application of the precise degree of stress to the cables.
There are two methods of pre-stressing concrete:

(a) **Pre-tensioning** (Figure 5.3)

Here the cables are stretched between the anchor blocks fixed to the pre-stressing bed. The framework is then arranged around the cables and concrete case. When the concrete has matured sufficiently, the cables are released and, in trying to return to their own lengths, they compress the concrete.

(b) **Post-tensioning** (Figure 5.4)

The method used here is that the cables are stressed after the concrete is set and has reached an adequate strength. The cables, or bars, are anchored at one end of the members and, using a special jack, are stretched until the right stress is reached and then anchored at the other end. The whole idea is to induce the concrete in the "tensile region" to be in compression. When a load is applied there remains, in the normally "tensile

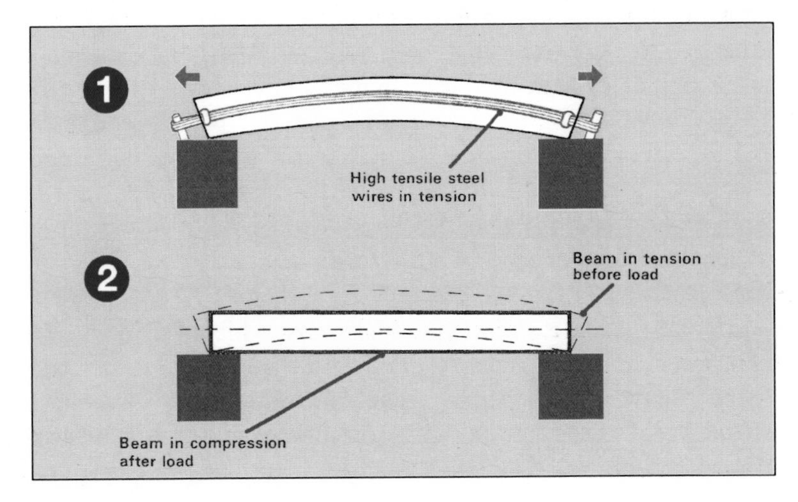

Figure 5.3 Diagram showing pre-stressing of concrete beams. (1) Pre-compression induced in the 'fibres' where under working load tensile stresses would be expected. (2) When the load is applied, there remains in the normally 'tensile zone' sufficient compressive stress to neutralise the tensile stresses of the applied load.

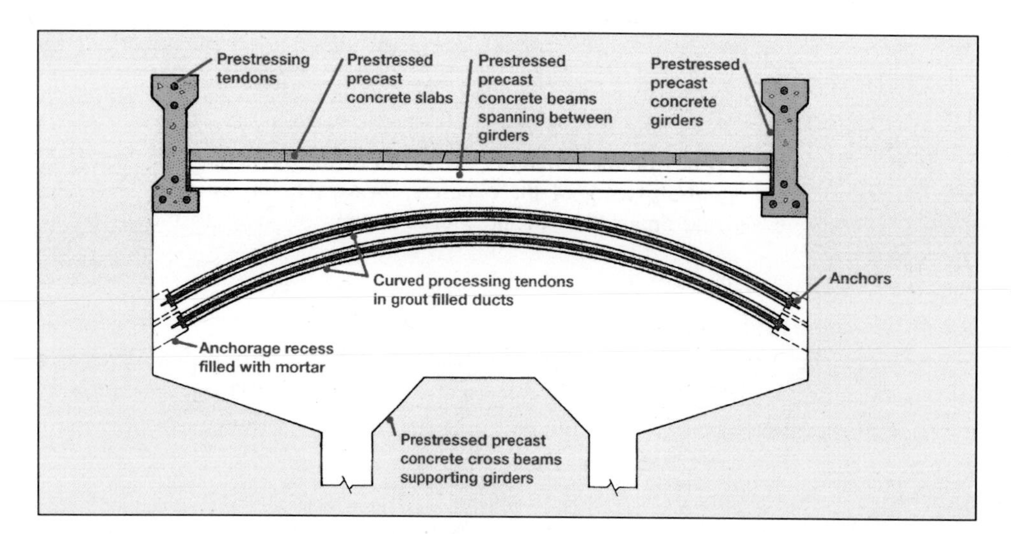

Figure 5.4 A typical example of the structured use of prestressed concrete.

zone", sufficient compressive strength to neutralise the tension.

5.1.7 Cast-iron

Although beams are no longer made of cast-iron there are many which still exist. A feature of all cast-iron beams is a large bottom flange, the top flange being smaller or, occasionally, omitted altogether. Stiffeners are cast on the web and the ends shaped to fit the head of the cast-iron column to which they are bolted.

5.1.8 Steel

Beams of structural steel are usually referred to by the function they are required to perform e.g. main beam, tie beam, joist. The term "universal beam" is used to denote one of a range of sections usually of the same type but varying in size and mass per kilogram per metre run. Most beams are of a rolled "I" section with, if necessary, additional flats riveted to the top and bottom flanges to give it added strength. Where fire resistance is required the same methods as specified for columns apply.

5.1.9 Steel lattice joists

This type of beam (Figure 5.5) consists of cold-rolled steel bars or tubes welded to top and bottom plates most of which are profiled for a particular reason e.g. the type of decking which can be used as permanent shuttering for a reinforced concrete slab. These beams are light but strong, easily erected and ideal for lightweight roofs or floors and are found in large single and two-storey industrial storage and commercial buildings.

5.1.10 Open-web beam – castella type

This type of beam (Figure 5.6) has been cut along a castellated line and then welded back together again. This increases the depth of the beam one and half times and reduces the deflection under load. Both steel lattice joists and castellated beams are often used to support ceilings as their design allows all types of services to be run through the beams.

5.2 Columns

A column is a structural member, which transmits a compressive force along a straight path in the direction of the member. Columns are usually thought of as being vertical, but any structural member, which is comparatively loaded, is governed by the laws of columns, despite its attitude.

The function of a column is to carry part of the weight of the building where an internal wall would interfere with the designed use or where a large open space is needed. A column is often designed to withstand only vertical loads and any eccentric loading greatly increases the stress.

Columns by themselves are often used for monuments. Non-vertical columns are often called by other names, such as struts or rakers, which are diagonal columns, which brace foundation piling. A bent is a line of columns in any direction. If a line of columns is specially braced to resist wind,

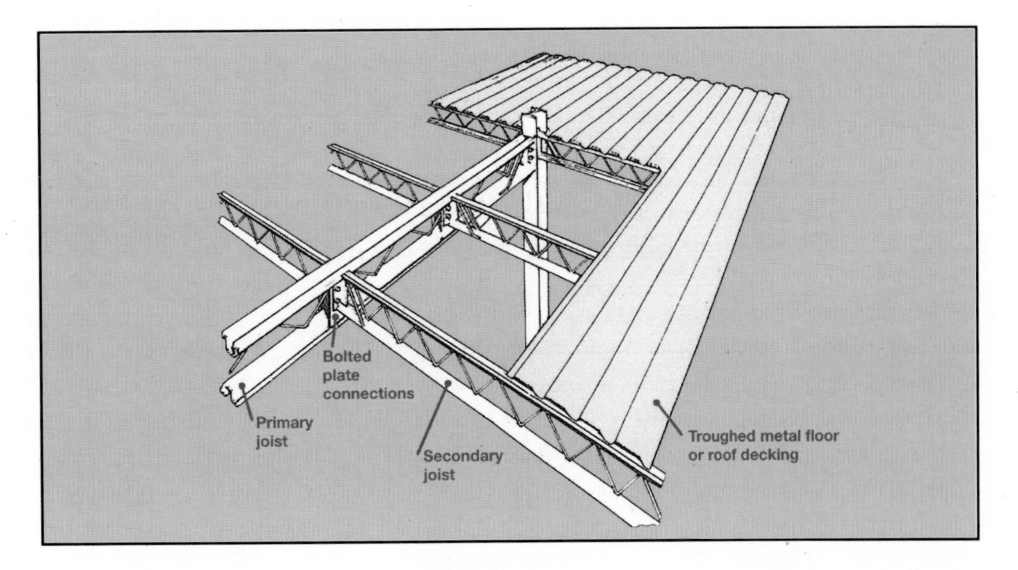

Figure 5.5 Typical construction for joisted floor or roof.

it is called a wind-bent. A bay is the floor area between any two bents.

A pillar is a free standing masonry load-carrying column, as in a cathedral.

In very old buildings, timber columns may be encountered. These may comprise simply smoothed off tree trunks. Large wooden columns may be ornamental as well as structural and may be hollow. The column itself may comprise curved, usually tongue-and-grooved sections glued together.

Where steel beams are I-shaped, steel columns are H-shaped, box-shaped or cylindrical. Beams are shaped like the letter "I" because the depth determines strength. Columns are shaped like the letter "H" and of a dimension that permits a circle to be inscribed through the four points of the "H". "I"-shaped steel sections may be used as a column, but it is wrong to speak of "I" beam columns.

5.3 Types of columns

5.3.1 Timber

At the beginning of the 19th century timber was normally used for columns in multi-storey factories and mills and some of these buildings still exist. Timber columns are usually found fitted with cast-iron caps, which accommodate the ends of the wooden beams. When these columns are located one above the other, on various floors, a cast-iron pintle (a bar of round section) runs through the beam in line with the column and transmits the load. This avoids the undue crushing force on the intervening timberwork.

5.3.2 Laminated timber

Techniques for laminating sections of timber are well established and usually replace the more costly balks of timber. The sections are described as "glulam" and their designed load-carrying capacity can be accurately calculated as well as their fire-resistance.

5.3.3 Brick

Brick columns are usually found in basements supporting beams which, in turn, take the load of the building above. A development of brick column building is post-tensioned columns. These are stressed in a similar way to post-tensioned concrete and resist forces, which would tend to overturn the column, by compressing the column

Figure 5.6 A typical open-web steel beam.

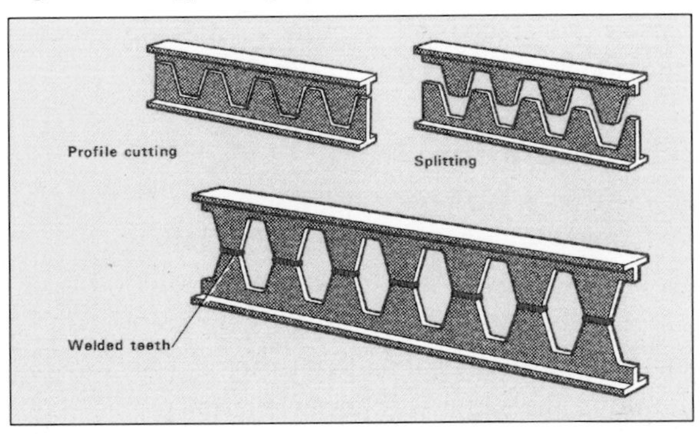

lengthways. These can be found in all areas of a building.

5.3.4 Stone

Stone columns in old buildings, even if particularly massive in appearance may not be as solid as they look. Some will have been constructed with facing stones filled behind with rubble and mortar. This rubble could have settled leaving the column load carried by the facing stones only. More modern columns may, or may not, be loadbearing but where they are they will most probably conceal steel stanchions behind a stone face. Firefighters should also be aware of the trend nowadays of erecting "stone" columns, which are made of tough plastic or fibreglass covering a steel girder.

5.3.5 Reinforced concrete

The reinforcing steelworks in structural concrete has developed to a very high standard today. Precast factory constructed units are probably more in use on sites using modular building methods but a lot of steel fabrication is done on site and the concrete poured into form work around the steel reinforcement. The fire resistance of a concrete column depends on:

(a) The applied load;
(b) The type and strength of the concrete;
(c) The dimensions of the column;
(d) The method of reinforcement;
(e) Its resistance to collapse.

A column should have at least the fire resistance of the elements of structure which it supports or carries and this, under the Building Regulations in England and Wales, depends on what type of structure they are a part of. Obviously a higher standard of fire resistance will require greater dimensions and adequate protection of the steelworks.

5.3.6 Cast-iron

Although seldom used in modern buildings cast-iron columns will be found in many old buildings and especially manufactories. There are numerous shapes and fittings and often different types will be found in the same building where extensions have previously been carried out. A common type (Figure 5.7) is a circular tube with a rectangular capping which carries the ends of the beams. The height of the column is the same as that of the floor and can vary from 2.7m to 6m whilst diameters can be 450mm on the lowest floor of a large building to 150mm on the top floor of a small building. The bases of the columns on each floor fit into the caps of the columns below but are not, usually, bolted to them. Quite often a central spigot fits into a socket and these can be either of wood or metal. It is not unusual to find cast-iron columns still standing after a fierce fire when the remainder of the building has collapsed.

5.3.7 Structural steel

Steel columns are usually of "I" section rolled as a single piece. Occasionally, where necessary they may be strengthened by flat plates riveted to the flanges and they can run up through more than one floor. The horizontal joists carrying the various floors would then be bolted or riveted to the column. Structural steelwork has the disadvantage of being unable to withstand the high temperatures generated under fire conditions and it will quickly loose its strength, buckle and fail. It must, therefore, usually be protected where fire resistance is required and the type of protection can be either "solid" or "hollow".

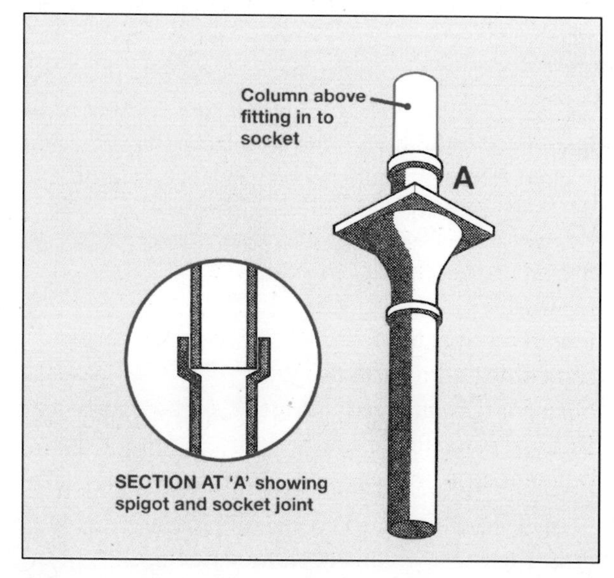

Figure 5.7 Sketch showing the construction of a cast-iron column and the method which is used to transmit the load through a floor.

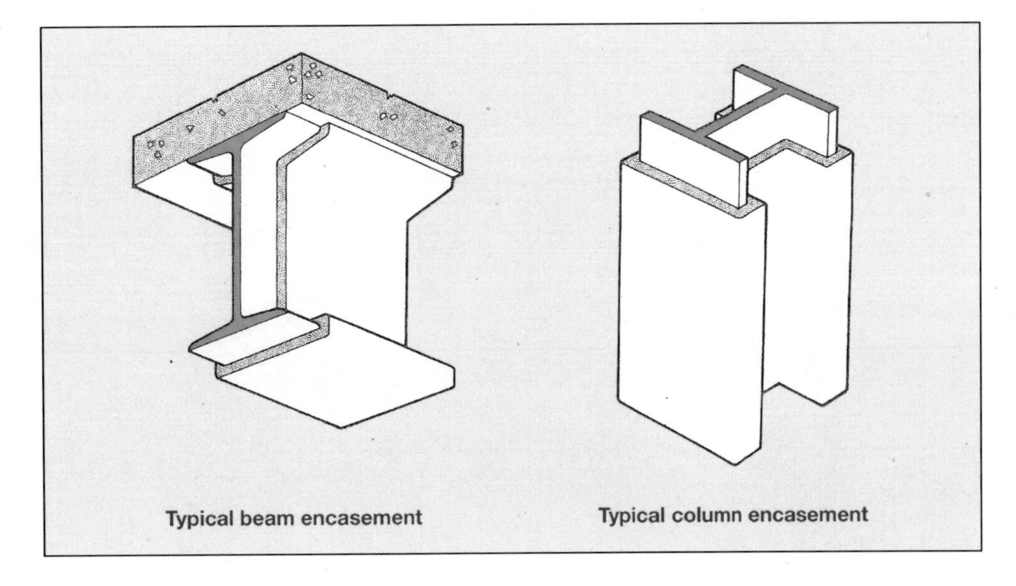

Figure 5.8 Example of "solid protection" to a steel section beam and column.

5.3.8 Solid protection

Nowadays this is achieved by (Figure 5.8):

(a) Concrete encasement; or
(b) Spraying with different types of mineral fibre vermiculite cements, magnesium oxychlorides etc., or
(c) Application, either by spray or brush, of intumescent paints.

The degree of fire resistance required in the case of (a) and (b) will depend on the density of the application and its thickness. The same applies in part to (c) although the chemical ingredients will dictate the amount of intumescing that takes place and also the protection afforded by the carbonaceous char.

5.3.9 Hollow protection

This is the encasement of steelworks (Figure 5.9) by fire-resistant boards and the Building Regulations (England and Wales) give guidance in the Approved Document B. Again the materials from which the boards are made vary from manufacturer to manufacturer. Vermiculite is often used, sometimes combined with gypsum, but there are other ingredients. Modern boards have gone away from asbestos for health reasons but these will still be found and give good protection.

A method now used is covered mesh protection, which combines fire-resistant compounds sprayed onto a fire-resisting metal mesh surrounding the steelworks. A third method is to fill in the hollow protection with additional thermal protection – mineral fibre, fibreglass, rock wool, foamed slag etc., all of which will add to the fire resistance of the steel. Needless to say most of these coverings should be finished off very carefully at joints, especially at floors and walls, to ensure that the whole is up to the required standard.

The fire resistance of the various methods of protection must be at least that laid down by the Building Regulations in England and Wales, for that particular type of occupancy. In addition any column must have the fire resistance of not less than the period required for any element which it carries and, if it forms part of more than one building or compartment, must comply with the maximum fire resistance for those buildings or compartments.

Heavy steel columns fail less readily than light ones as the thermal capacity i.e., the ability to absorb heat, of the heavy column is greater for the same temperature rise. Consequently, a light steel column will probably require more protection than a heavy one.

Columns can also be differentiated by the manner is which they generally fail:

(a) Piers – are short squat columns, which fail by crushing;
(b) Long slender columns – fail by buckling. In buckling, the column assumes an "S" shape,
(c) Intermediate columns – can fail in either way.

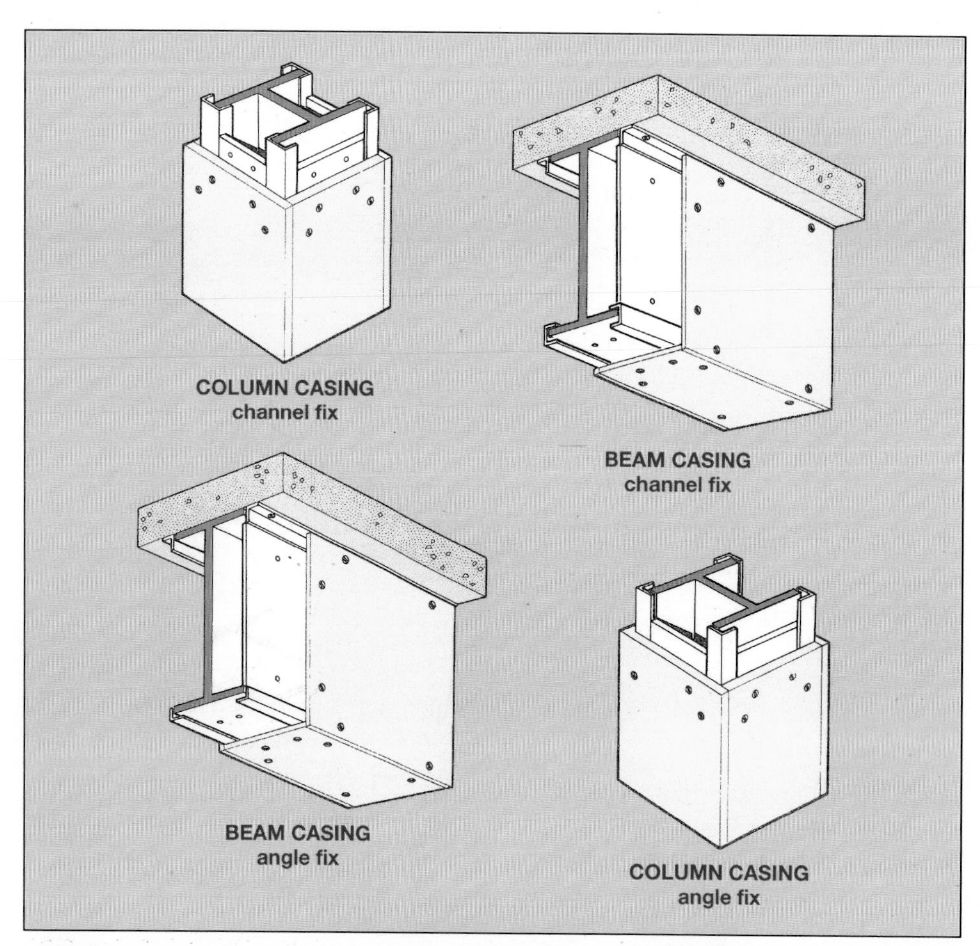

Figure 5.9 Different types of hollow protection to beam and column.

Figure 5.10 Type of concrete "plank" showing its final position right.
Photo: Essex Fire Brigade.

5.4 Floors

5.4.1 General

In all, except single-storey buildings, floors are a principal structural element and vary greatly according to the design of the structure. In a steel-framed building the frame is designed to support the floor and, therefore, a designer can use precast concrete slabs which will span between the joists (see Figure 5.10). In a reinforced concrete-frame building the whole frame and floors are usually poured in sequence to make a monolithic structure following which the shuttering is removed. A further development has come into use whereby the steel joists are spanned by light metal shuttering

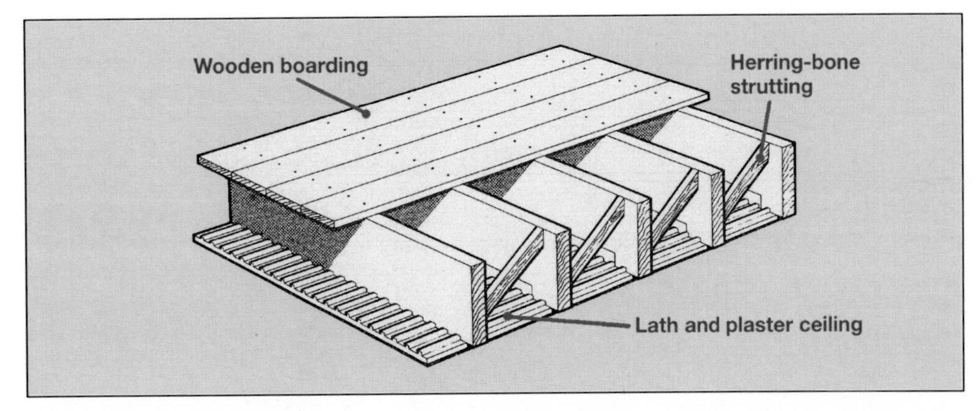

Figure 5.11 A typical timber-joisted floor as used in domestic houses.

onto which a concrete floor is poured. The shuttering is left in place and becomes part of the floor. These are three examples of possible types of floor for, perhaps one office block design.

Floors can be regarded as being composed of three parts:

(a) The actual loadbearing members;
(b) The upper surface or finish of the floor; and
(c) The lower surface of ceiling of the compartment below.

In all but the most basic structures, where one construction combines all three, all three parts will be separate and identifiable.

For example, in a timber floor in a small house the loadbearing members are the joists, the surface is the boarding and the ceiling is of plaster. Here the joists provide the preponderance of the strength of the combination and, while the boarding adds to the rigidity, it is not an essential contributor; the joists themselves support the plaster.

Compare this with a reinforced concrete floor in which all three parts, ceiling, floor surface and structure may be completely merged. The whole thickness of the concrete slab contributes to the strength of the floor and the upper and lower surfaces provide the floor and the ceiling. This factor becomes more important when we come to consider membrane or suspended ceilings in compartmentation.

Construction of the more common types of floor is explained in the following paragraphs.

5.4.2 Timber floors

Timber floors will be found in many types of building and, in most cases, are required to provide certain levels of fire resistance according to the type and size of the building. Most timber floors are underdrawn with ceilings of various materials and these usually add considerably to the fire resistance. Other factors in the performance of timber floors in fire include:

(a) Whether the flooring is plain edged (butted, tongued and grooved) or is chipboard or plywood;
(b) The thickness of the flooring;
(c) The loadbearing capacity of the joists (and the load imposed).

5.4.3 Timber-joisted floors

The timber-joisted floor (Figure 5.11) has been generally used for the upper floors of houses of all periods. Butt-jointed or tongued and grooved boarding between about 16 and 32mm thick is used, laid on wooden joists usually not less than 50mm thick and varying in depth from 128 to 180mm according to the distance spanned. These joists may be prevented from twisting by strutting, of which each unit may be either a solid board or two cross herringbone struts, although nailed boards will have the same effect. On the underside of the joists is the ceiling usually, in modern work, of building board with a thin coat of plaster. This leaves a space between each joist enclosed by the floorboards and ceiling which constitutes hazard because fire can travel, undetected, in it. In the case of a hearth fire, in particular, it is often necessary to lift the floorboards at intervals to verify that the fire is not travelling to some other part of the structure.

In Scotland the laths or plasterboard are nailed to small battens called "branders" which run across the underside of the floor joists (Figure 5.12). These branders prevent the joists from twisting

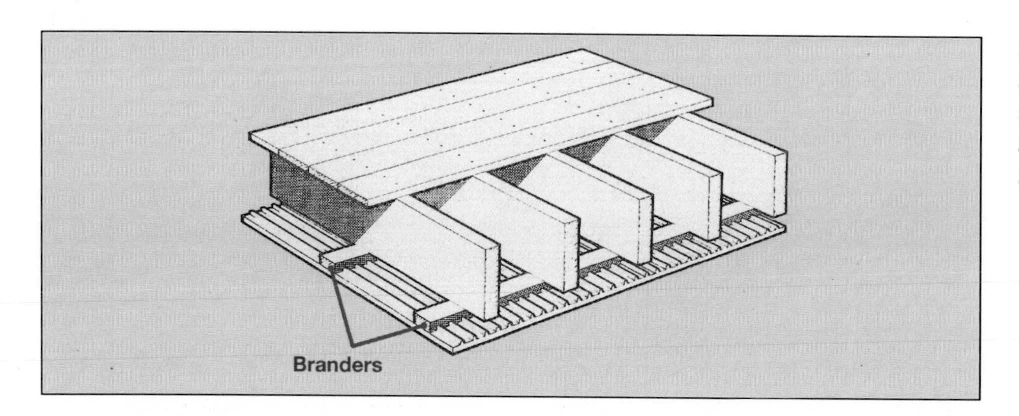

Figure 5.12 Method of securing the ceiling laths to branders, sometimes used in Scotland.

but, since the laths or plasterboard are held away from the joists fire can spread through the small air-space more rapidly than in other types.

5.4.4 Floor supports at walls

The way that the joists are supported on the walls is of importance to the firefighter and several methods are used.

In old work the joists are simply built into the wall (Figure 5.13(1)), and there is a risk that the collapse of the joists in a fire could lever the wall off balance. A more common method is the provision of a wood wall plate on to which the ends of the joists are nailed (Figure 5.13(2)).

This, if built into a wall, tends to weaken it. A third method is to build in a wrought steel wall plate (Figure 5.13(3)). Whichever design is adopted, unless sufficiently large "joist pockets" are allowed, collapsing joists will lever the wall off balance.

Of the more satisfactory methods used, from the point of view of fire, one is to support the wall plate on wrought-iron brackets built into the wall (Figure 5.13(4)), a second is to corbel the brick-work out to form a ledge for the wood wall plate (Figure 5.13(5)), and a third is to reduce the thick-ness of the wall by 114mm at each floor level and to rest the wall plate on the ledge (Figure 5.13(6)).

Many houses are floored with plywood or chip-board and this is laid in either one continuous sheet which could cover the whole of an upper floor or in large squares each about 900mm × 900mm. These are screwed to the joists and it is obvious

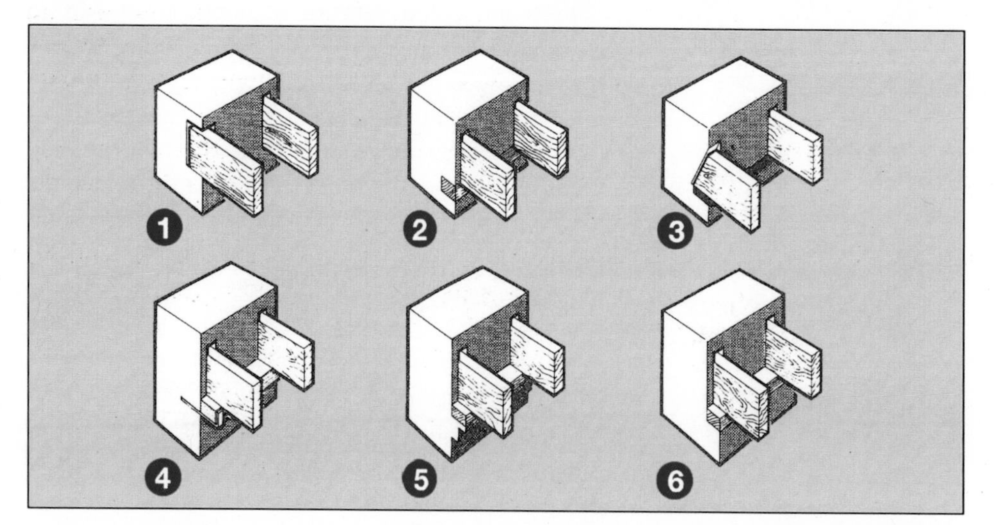

Figure 5.13 Sketches showing various arrangements for supporting floor joists: (1) joist with square end in pocket; (2) joist carried on wooden wall plate; (3) joist with splayed end in pocket; (4) joist carried on wooden wall plate carried on bracket; (5) joist carried on wooden wall plate on corbelled brickwork; (6) joist carried on wooden wall plate on a ledge formed by reducing the thickness of the wall.

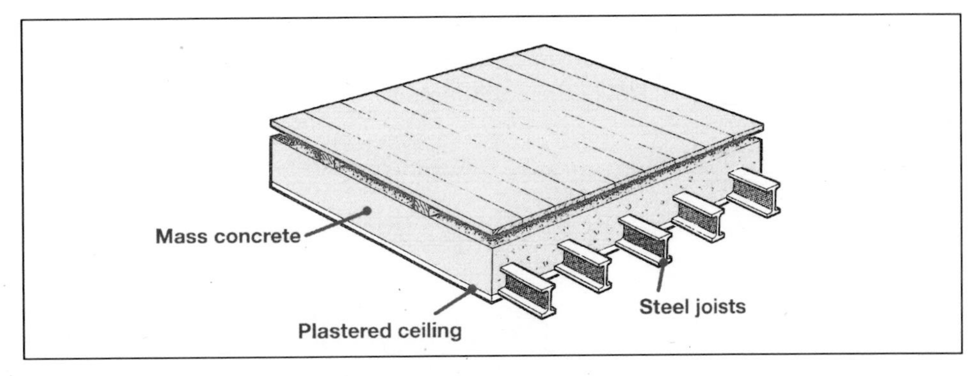

Figure 5.14 Mass concrete and steel filler joist floor

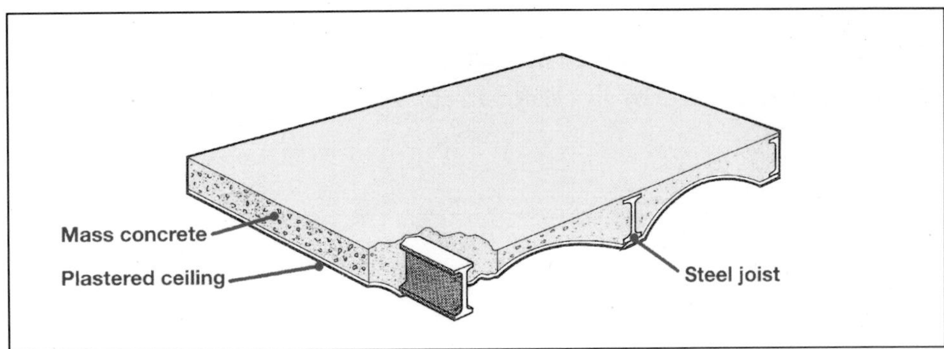

Figure 5.15 Steel filler joist with arched construction mass concrete floor

that they present a problem to firefighters seeking to inspect joists or floor voids for fire damage or fire travel.

For many years ground floors of houses and similar constructions have been of concrete but this is changing back to the old style of suspended timber flooring. The ground floor is now frequently constructed in a similar manner to the upper floors and services will be run under them as elsewhere. Occasionally the void will be underdrawn with polystyrene sheets or filled with mineral wool for insulation purposes.

5.4.5 Brick arches

This is a type of construction mainly found in old warehouses, mills and manufactories and they may be supported on brick piers, cast-iron columns and beams and even, occasionally, on huge timbers. The upper surface is often filled in with concrete to make a level and boarded or screeded over.

5.4.6 Steel "filler" joists and mass concrete

There are many varieties of this type of floor but the principle employed is to divide up the area to be filled in by steel joists set at intervals sufficiently small to be spanned by mass i.e. unreinforced, concrete.

Two types of floor are illustrated in Figures 5.14 and 5.15. In Figure 5.14 light steel joists, 100 × 45mm, are placed 300mm apart and the space between and above filled in with 200mm of concrete. The top is finished off smooth and may be boarded, tiled etc., as required whilst the underside is usually plastered. This type has a very good fire record especially when supported on substantial brick walls.

In the second type (Figure 5.15) the steel joists are heavier and spaced more widely and the concrete is arched up from the bottom flange of the joist to reduce the weight of the floor. The thickness of the concrete at the centre of the slabs may be as little as 75 to 100mm and, in a fire, there is a danger of the slabs cracking away from the steel. In both types the lower flange of the steel joist may be entirely exposed to the full heat of any fire from below.

5.4.7 Reinforced concrete (RC)

This type of flooring has developed from a fairly simple reinforcing steel rod to a highly sophisticated, interwoven steel mesh and rod combination onto which is poured a particular type of concrete and the strength and stresses on the whole can be very accurately calculated.

Plastic or metal moulds (Figure 5.16) can shape

the actual under-configuration of the floor. These are sometimes called "waffle" or "honeycomb" floors and Figure 5.16 illustrates the combination of steelwork and glass fibre moulds required formulating a floor of this kind.

Figure 5.17 illustrates a sectional view of the type of reinforcing steel found in an RC beam.

An older type is shown in Figure 5.18. The construction is lighter and resembles the "waffle" type except that it has continuous arches. It is composed of RC beams spaced between 450 to 610mm apart with the spaces spanned by RC slabs structurally continuous with the beams. The actual thickness of the floor could be as little as 50mm and it is, therefore, inferior in fire resistance to the heavier RC type.

5.4.8 Pre-stressed concrete

Here pre-cast pre-stressed, either hollow or solid concrete planks or sections are usually used to span between structural steel beams. After they are laid they are covered with concrete which bonds the planks together to make the finished solid floor. Two types are illustrated in Figure 5.19 and 5.20. Some are designed with in-built tie bars, which help in the bonding when the concrete is overlaid.

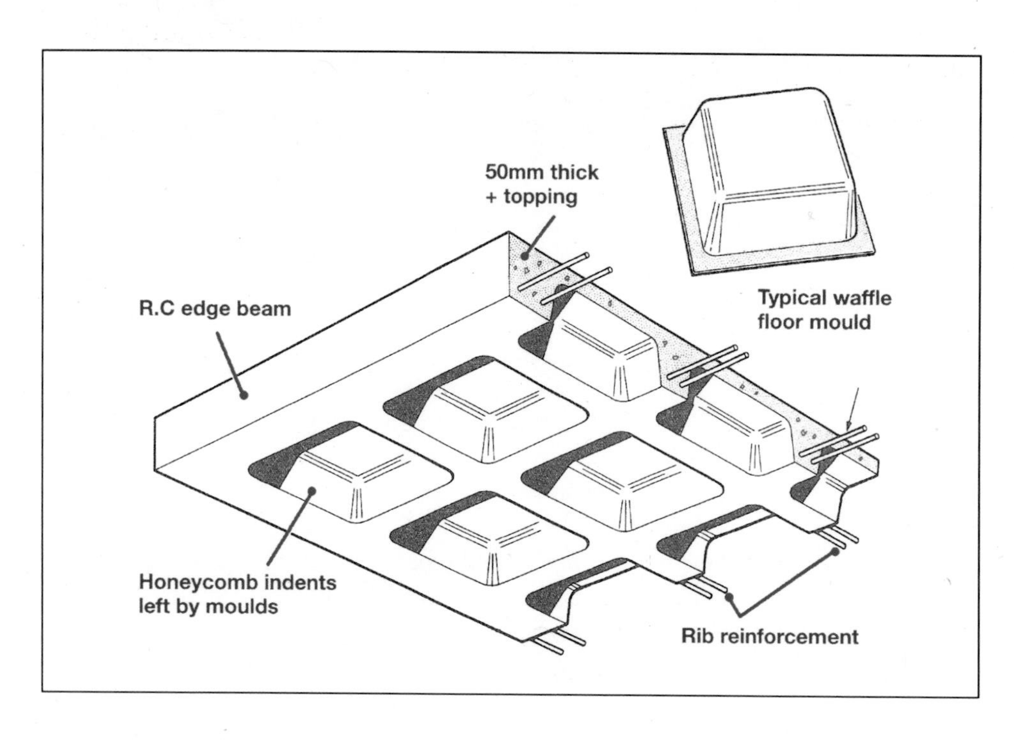

Figure 5.16 An example of a waffle or honeycomb floor.

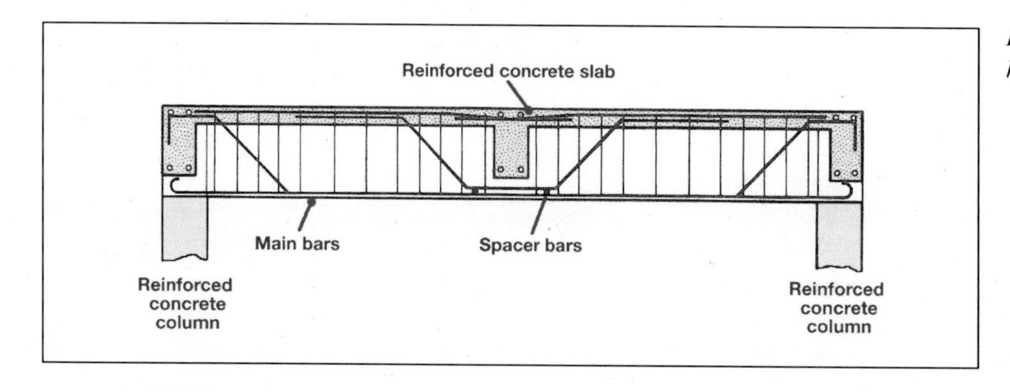

Figure 5.17 RC beam with heavy reinforcement.

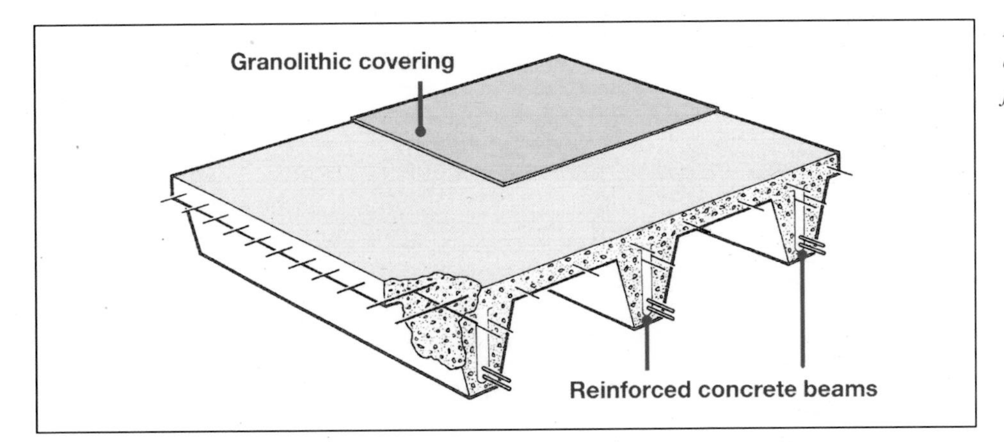

Figure 5.18 Reinforced concrete rib and panel floor.

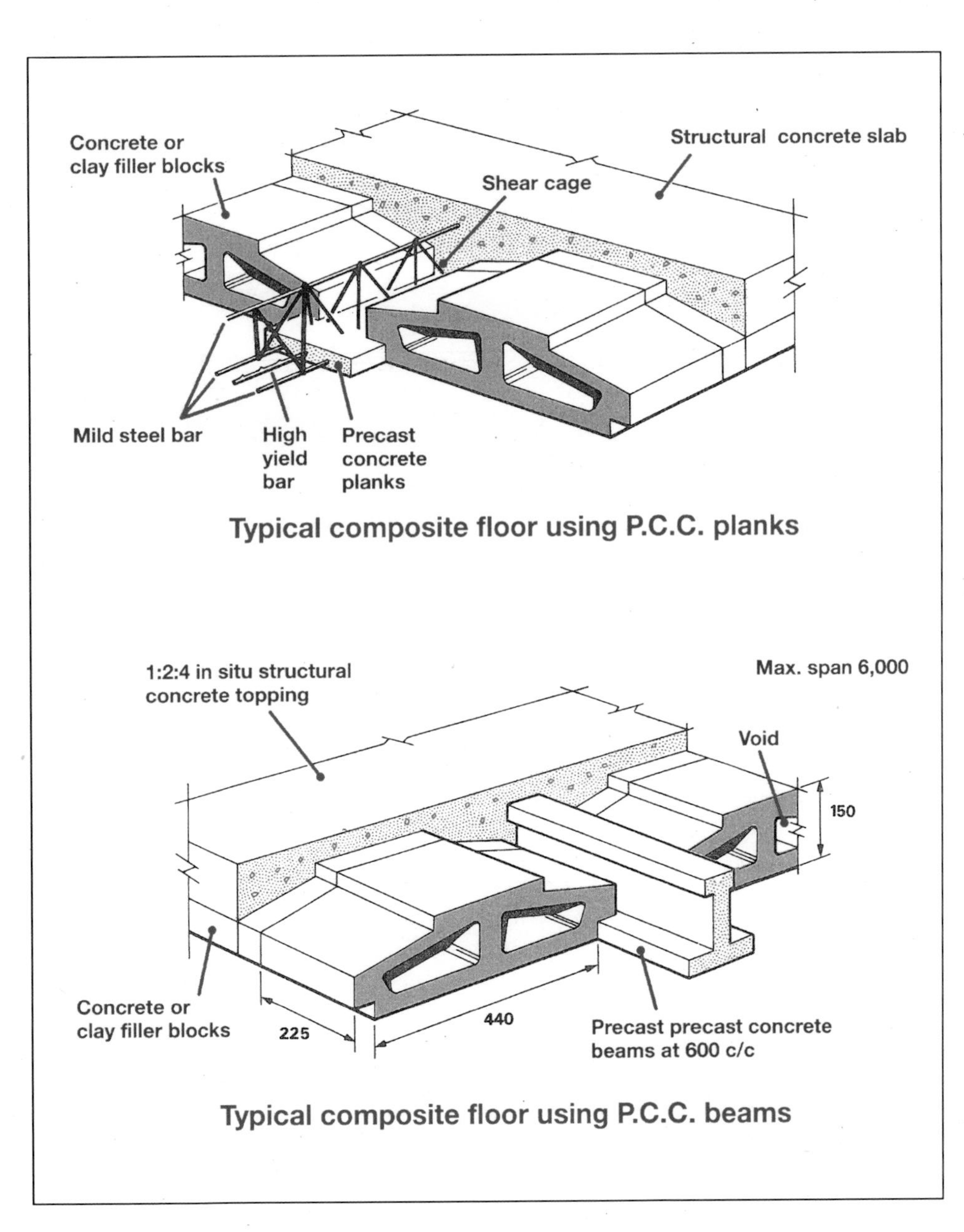

Figure 5.19 Two examples of plank floors overlaid with concrete.

forcing steel, depending on the proposed loading on the whole floor (see Figure 5.21).

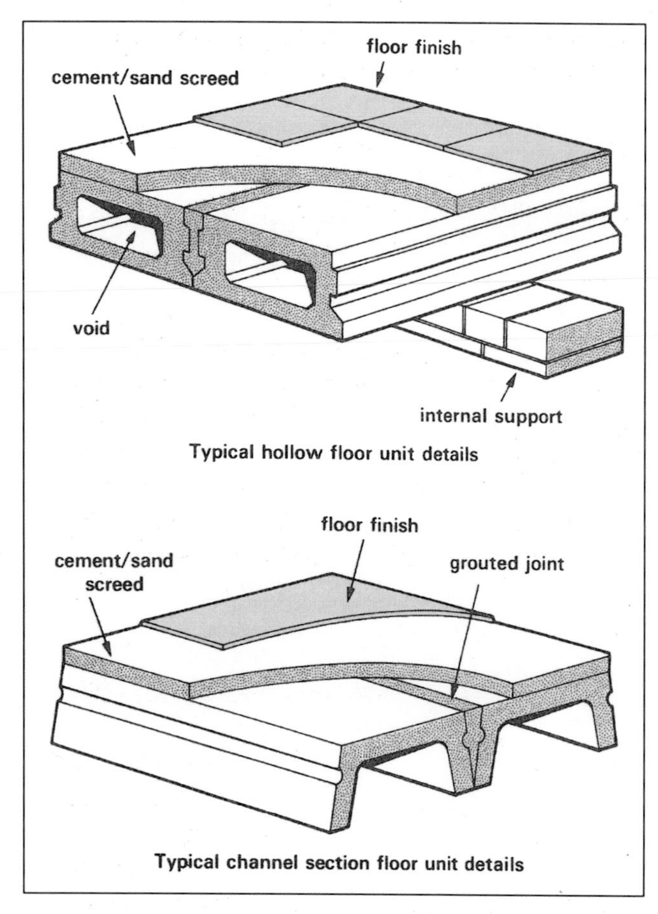

Figure 5.20 Unreinforced plank floors with light concrete topping.

5.4.9 Hollow block and plank

There are numerous variations of the types of concrete planks, which can be quickly laid with spanning steel or concrete beams. Examples of these are shown in Figure 5.20. Another lighter variation consists of hollow clay floor blocks held together in a light concrete topping with, or without, reinforcing steel, depending on the proposed loading on the whole floor (see Figure 5.21).

This type of floor has a good fire record although a fierce fire will tend to spall off the lower edge of the tiles. The remainder of the construction, however, is usually sufficient to maintain the stability of the floor.

5.5 Roofs

5.5.1 General

It has been stated that "a roof is a structure, which surmounts a building to keep out the weather and may be flat, pitched or curved". Since that statement was made there have been significant developments and advancement in design. Whether this statement could still be applied to some modern buildings with roofs made of fabric, glass, plastics, tubing, cables etc., and which are suspended, cantilevered, sometimes geodetic, frequently braced and, occasionally, inflated is problematical.

Some of the materials now used, such as polytetrafluoroethlyene (PTFE) have a design life of many years (25 years or more) and the use of this technology appears to be finding increased applications. This is very much the case where there is a reluctance to destroy many fine buildings in city centres to provide a protected environment for shoppers. The installation of a lightweight fabric roofing structure may find increasing favour in the future.

PTFE-based roofing materials may present problems if subjected to heating in moderate-sized fires, but the toxic potency of products from such fires

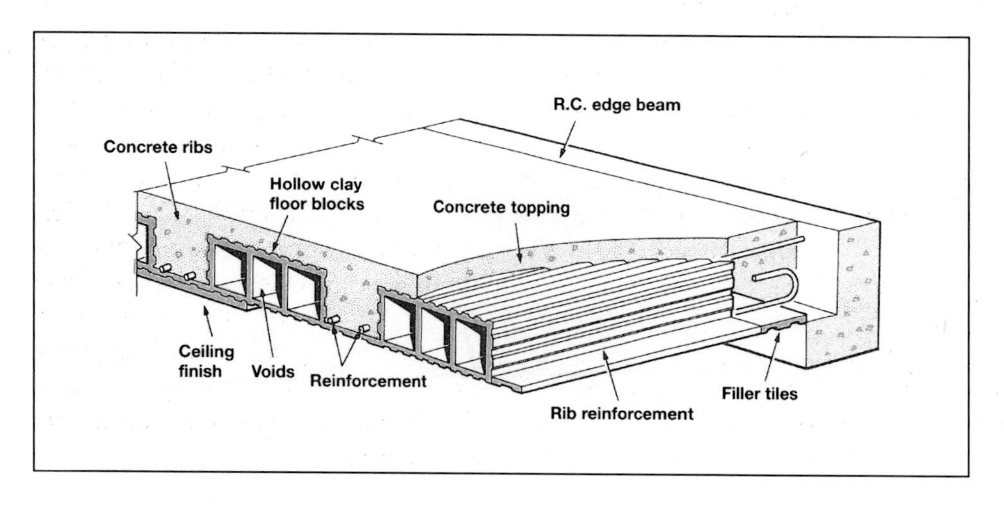

Figure 5.21 A hollow clay block flow with reinforcement and a concrete topping.

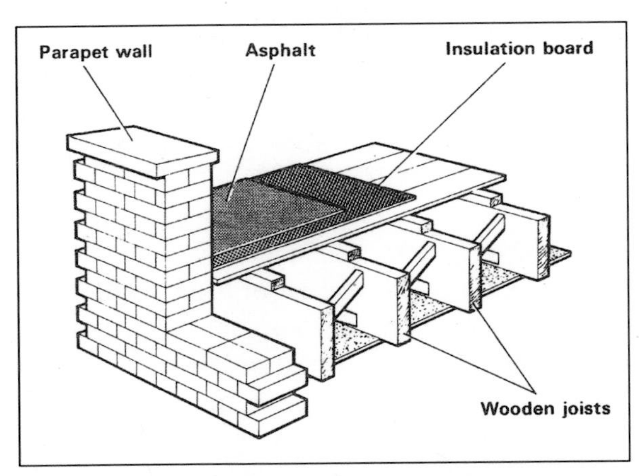

Figure 5.22 Wooden-joisted flat roof with parapet. For simplicity details of the damp courses have been omitted from the parapet wall.

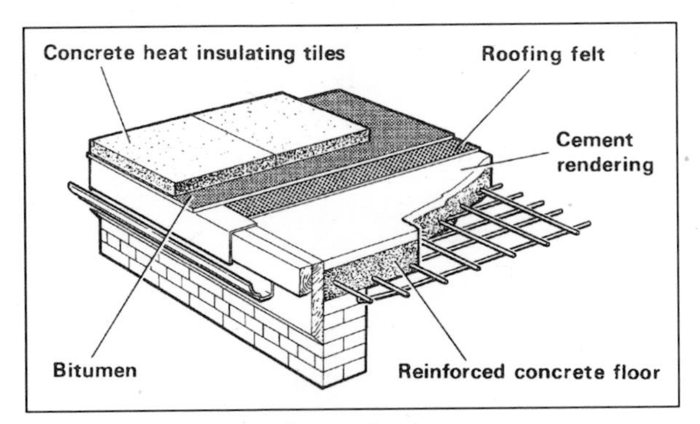

Figure 5.23 Concrete flat roof without parapet.

could in most circumstances be no worse than many other materials. Under certain conditions, products are evolved when PTFE or other fluoropolymers are subjected to a specific set of thermal decomposition conditions. They have a toxic potency to rates up to 1000 times greater than the combustion products of wood or other common materials.

Whether or not this will make a significant contribution to the overall toxic potency of the fire effluent will depend upon the size and nature of the original fire and the mass of PTFE-based material decomposed. It is therefore advisable to place roofing material well above potential fuel sources, and to avoid using such materials near the edge of roofing structures.

Roofing panels of PTFE-based material are commonly welded together. It is current good practice to design the welds to fail and vent the fire at low temperatures. However as an additional precaution, the panels in tension-supported structures should be designed such that they cannot become detached at one end and hang down into the fire.

Eythlene/tetrafluoroethylene (ETFE) glass fibre roofing materials and other materials have not been found to produce significant toxic products.

The design of many single-storey industrial buildings appear to carry the walls straight over into the roof and consist of polymerised insulation sandwiched between metal cladding. Others include large areas of glass or polycarbonate, which, again, cover both walls and roof. This section will, however, describe some of the commonest and simplest types of roof but emphasises the need for firefighters to note the construction of new types of roof in their areas.

5.5.2 Flat roofs

The construction of flat roofs varies from what is little more than a weatherproofed floor to a concrete plank assembly which is, itself, weatherproofed by screeding, grouting and bitumenising. Figure 5.22 shows a traditional construction like an upper floor, covered in a fibreboard and bituminised felt or poured asphalt. Figure 5.23 illustrates a floor of reinforced concrete with a bituminised weatherproofing and heat insulating tiles. Both are common basic constructions and the variety is endless.

5.5.3 Pitched roofs

Probably the most common type of roof used in 90% of houses and residential accommodation is the pitched roof.

5.5.4 Close-coupled roofs

The simplest and most common form of pitched roof is called a close-coupled roof. Timber rafters about 300mm apart run from a ridge-board to a wall plate on the top of an external wall at eaves level. These rafters carry the sloping roof itself, which can be of various materials. Where rafters are longer than about 2m they require support in the middle (Figure 5.24). This is done by a timber called a purlin, which supports the centres of all the rafters, and, on long spans, is itself supported by timber struts placed at intervals on the tops of internal walls. On every large close-coupled roof there may be two purlins under each slope.

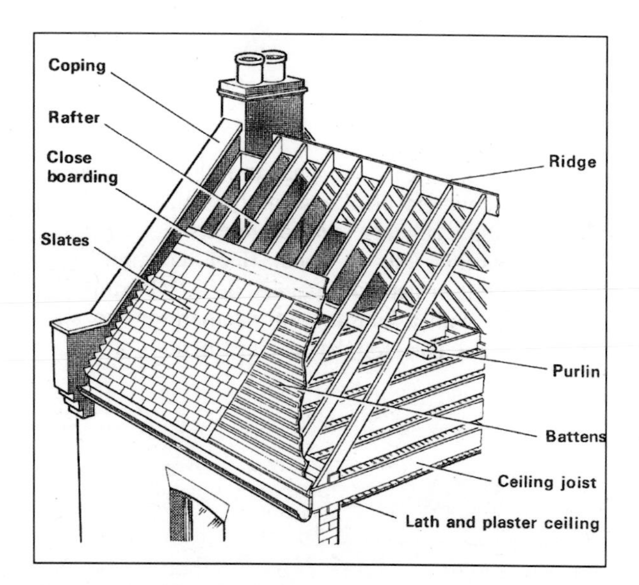

Figure 5.24 Sketch showing the more important features of the form of roof construction often used in domestic dwellings.

The ceiling joists are nailed to the wallplate and to the rafters (metal connectors may be used) so that the whole construction forms a series of triangles, which resist the tendency for the roof to spread outwards (see Figure 5.25). Occasionally the two sets of rafters are also connected together at about half the height of the roof by transverse timbers called collars. These help to stiffen the roof by reducing the free span of the rafters.

The rafters may continue beyond the line of the wall and overhang to form the eaves. Horizontal boarding may often be used below the eaves to keep out draughts, birds etc., from the roof space. In addition the spaces between the rafters should be filled with bricks to roof level but this is not often done.

This type of construction is mostly used in houses as well as for great numbers of other small buildings and there are several variations. For instance, the ceiling joists may be set up above the level of the top of the walls so that the rafters form part of the ceiling in the room below.

This is known as a "camp roof". Again, it may be found that rooms have been formed in the roof space if it is high enough by means of vertical framing (see Figure 5.26), the whole being lined with some sort of wallboard. The framing thus forms vertical walls to the rooms and there is some sloping ceiling on the underside of the rafters and a horizontal portion on the underside of the collars.

5.5.5 Mansard roofs

A mansard roof is a special type of pitched roof and instead of the roof running up at a constant angle from eaves to ridge there are two angles. One is a very steep pitch from the eaves to room height (Figure 5.27(1)) and then a flatter pitch above (Figure 5.27(2)). The object of this is to enable a room to be built inside the roof space, which, in effect, becomes another storey.

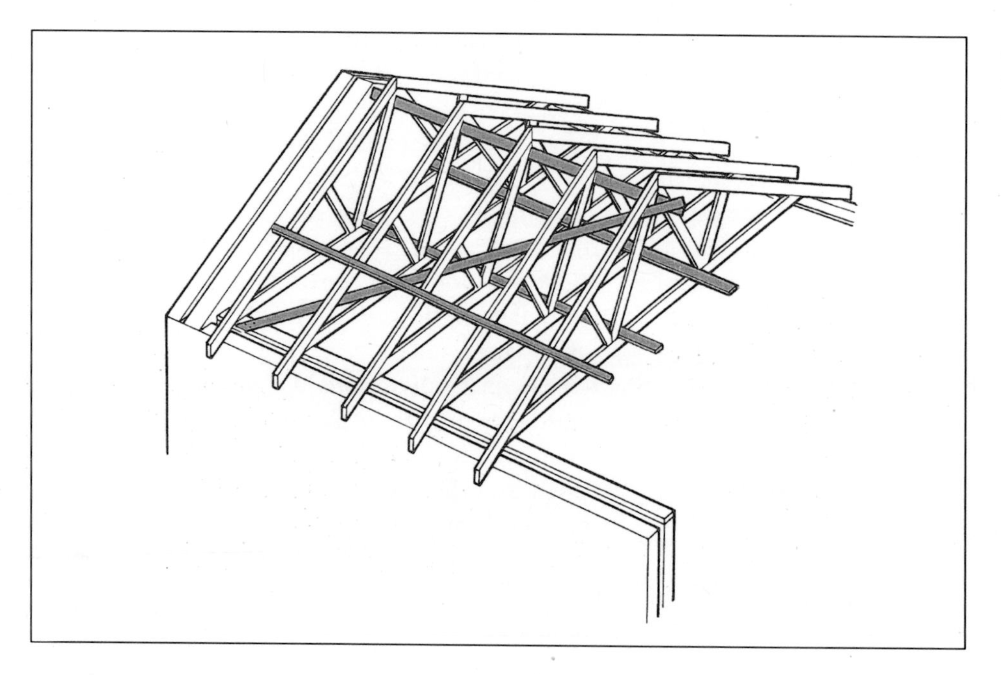

Figure 5.25 One method of joining and bracing roof members.

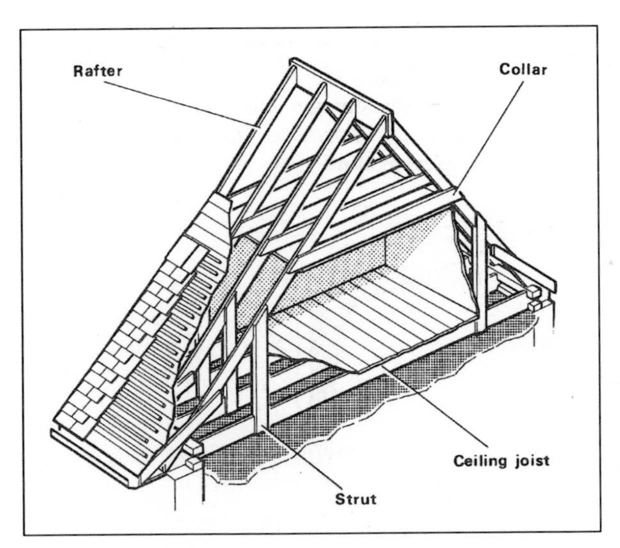

Figure 5.26 Sketch showing how part of the roof void may be utilised to form a room.

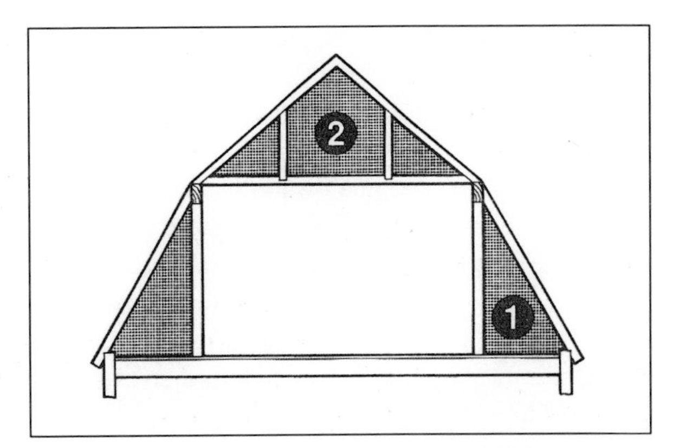

Figure 5.27 Diagram showing the voids left in mansard roof construction: (1) is the void left by the steep pitch and (2) the void left by the flatter pitch.

5.5.6 Trussed roofs

This type of roof is used to spread the load and direct it to the walls and the ground. Construction varies widely both in its use and design. The older styles of timber rafters, either tensioned by iron rods or timber, and the cast-iron trussing found in large old mills and warehouses etc., have been superseded by laminated timber, steel tubing, steel lattice girders and aluminium. The large spans over concert halls theatres and cinemas are types of trussed roof as are the ultra modern tubular steel and geodetic fabrications in shopping centres and other large enclosed areas etc.

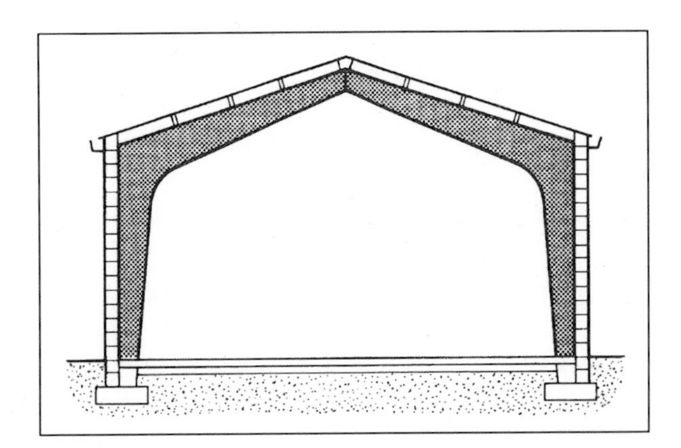

Figure 5.28 Typical Portal or rigid frame roof construction.

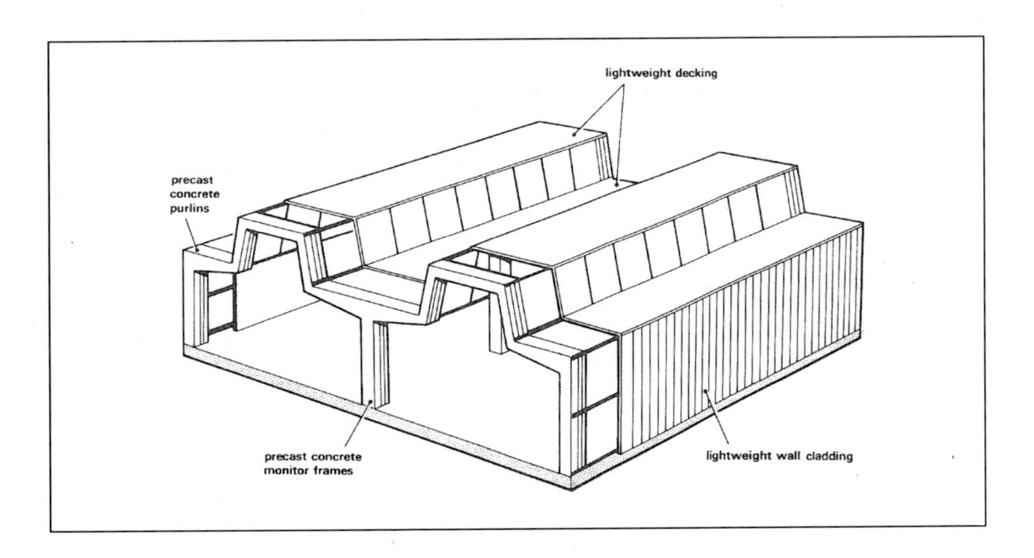

Figure 5.29 Typical monitor roof profiles.

5.5.7 The Portal or rigid frame roof

The Portal frame consists of, essentially, a continuous member conforming to the outline of the roof and connected to vertical columns. This continuous frame has the effect of passing the roof loading to the rest of the structure (Figure 5.28). They are especially suitable for single storey industrial or storage buildings giving long, wide open areas. Steel, aluminium and laminated timber are used but most are of precast prestressed concrete units.

5.5.8 Monitor roofs

A type of roof found in factories or stores is illustrated in Figure 5.29. It is of a relatively light weight and is designed to give the maximum amount of light by the use of "upstands" of glass or polycarbonate. The nonglazed portions are usually light decking and the walls are generally of a lightweight "sandwich" cladding, the whole supported by precast concrete frames.

5.6 Roofing materials

5.6.1 Slates and tiles

The simplest form of pitched roof covering consists of slates or tiles nailed or laid on wooden battens which are, themselves, nailed to the rafters. Felt is sometimes laid under the battens for the purpose of heat insulation or weatherproofing. In better quality work, boarding and felting may be employed; the tiling battens being nailed through them to the rafters below. Slates may be thin and comparatively light in weight or they may be thick and heavy. All slates must be nailed on.

Plain tiles are flat or slightly curved on both sides and have "nibs" at the top, which are used to hook them onto the battens. Tiles rest in position by their own weight and in only the best work are they secured by nails to the battens. Pantiles are heavier and curly in shape and are hooked on in most cases. Concrete tiles will also be found to resemble both the interlocking and the plain clay tiles.

To prevent entry of rain, especially where there is a heavy weather exposure, slates and tiles may be bedded in mortar. This is a process known as "torching".

Tiles can also be found made of asbestos-cement and are extensively used in bungalows, sports pavilions and other inexpensive structures. They can be distinguished by the considerably larger area covered by each tile and, owing to their light weight, should require less heavy roof timbers.

5.6.2 Sheeting

Corrugated iron, aluminium and corrugated asbestos cement sheet require a different form of roof construction because the large sheets of material are fixed to purlins not battens. With these materials no common rafters are used. The principal rafters (or trusses in large buildings) are widely spaced – as much as 2–3.5m apart in the case of principal rafters and 3-5m in the case of trusses – and themselves support the purlins. There is a line of purlins to each row of corrugated sheet placed under the line where the sheets lap over one another.

Roofs of corrugated iron and corrugated asbestos cement sheet have poor insulation and are often underdrawn by matchboarding or plasterboard. They will take little weight and firefighters must exercise the utmost caution when working on or near them.

5.6.3 Cladding

Nowadays many single-storey industrial buildings are clad in metal sheeting, which is corrugated but usually in square or oblong section. As stated before, often the roof is all one part with the wall (see Figure 5.30) and consists of a polyisocyanurate or polyurethane foam insulation sandwiched between two sheets of steel sheeting or aluminium-alloy sheeting. These are fixed to metal rails along the walls and metal purlins in the roof.

This type of metal cladding is increasingly being used for finishing the roofs of commercial office buildings, leisure centres, schools etc., and is fixed to an appropriate metal or timber frame. The fixings vary widely and, in some cases, are extremely complicated to compensate for expansion, water penetration, building movement, wind etc.

The metals used e.g. steel, aluminium, copper, lead, zinc are alloys depending on what type of

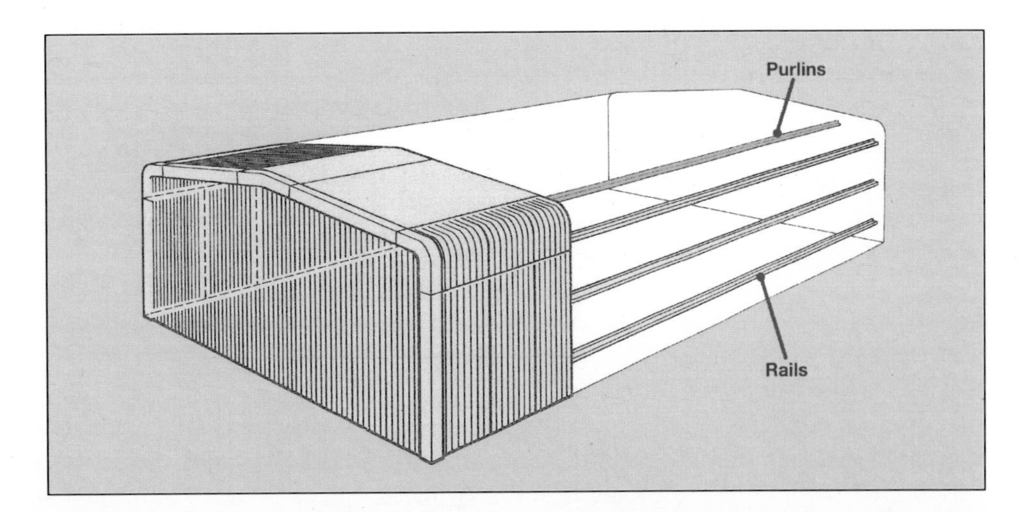

Figure 5.30 Sketch illustrating typical framing and metal cladding.

Figure 5.31 An example of metal cladding designed to be a particular feature of a roof. Photo: Broderick Structures.

"finish" is required. The metal can be shaped over practically any form as illustrated in Figure 5.31.

5.6.4 Decking

Decking is usually found in flat or nearly flat roofs. The supports can be steel or aluminium girders or tubing, timber or reinforced concrete. The decking can be made up of almost any type of board e.g. fibreboard, strawboard and weatherproofed with layers of asphalt or bitumenised roofing felt, possibly heat sealed and topped off with a further protective layer of tiles or heat reflecting material. Generally these roofs are safe for firefighters to work on but that will depend on the size of the fire beneath them and the strength of the roof supports.

5.7 Behaviour of roofs and roofing material in fire

5.7.1 General

Roof coverings are, in general, non-combustible or, at least, not readily combustible (an exception being thatch) so a roof is not normally vulnerable to fire from an external source. Generally it is the way roofs are built rather than the material used that causes difficulties for firefighters. The pitched roof presents problems because of the large unused spaces that exist i.e., lofts, attics, voids etc., between the ceiling of the rooms below and the weather covering. As stated elsewhere these voids can extend, unbroken, over several dwellings or, in some cases, over the whole of the building. The amount of timber present, often of light cross-sectional area, the rising heat and smoke from a fire in the building coupled with the fact that there is seldom easy access to these areas from below or through the roof covering, means spread of fire into the void presents a difficult and punishing period of firefighting.

5.7.2 Fireblocking

Most modern residential buildings with this type of roof are fireblocked, i.e., where required in construction, fire-resisting foam blocks be inserted to prevent fire spreading through cavities up into the roof. The roofs are also compartmented and precautions are taken to stop fire-spread from one void to another including over the top of the compartment division.

5.7.3 Connectors

As mentioned above, a modern trend is for the light timber in roofs to be put together with metal connectors. These have been known to expand and fall out in a fire and leave the roof unsupported with the certainty of roof collapse.

5.7.4 Slates and tiles

A fire attacking the underside of a pitched roof can release slates or tiles and these can slide off causing injuries to firefighters below. The rate at which this can happen will be dictated by whether the roof is underdrawn with closeboarding or only with felt.

5.7.5 Steelworks

Steelworks or wrought iron, especially of light section, is vulnerable to a fierce fire and a fairly rapid collapse can follow when the steelworks is unprotected. A lot will depend on the type of roof it is supporting and whether there is any roof venting e.g. automatic vents, thermoplastic rooflights.

5.7.6 Cast-iron

Cast-iron is of greater cross-section and density than steel and such trusses can still be found in place after a fire providing there has not been a sharp change in temperature i.e., a heated truss struck by a jet of water.

5.7.7 Trussing and cladding

The complicated trussing in, for instance, a cinema or theatre roof is usually of light section. Although protected by the auditorium ceiling firefighters must be aware of the possibility of a complete collapse in a large fire. The reaction of insulated metal-clad roofs will depend largely on the types of support and the fixings. Both will be of light construction and, again, venting of the roof will keep the temperature down and the roof up.

5.7.8 Concrete

Concrete is usually reinforced in some way and providing the steelworks is not exposed quickly to the fire e.g. by spalling of the concrete cover, it can still maintain its support.

5.8 Rooflights

A rooflight is a form of window in the plane of a roof and fixed. An opening rooflight is referred to as a 'skylight'. Rooflights are placed in buildings primarily to allow natural light in; some are fitted with louvered panels for ventilation. Many conventional rooflights still exist for example: lantern lights which consist of vertical glazed sides and a glazed roof, the sides are often arranged to open or are fitted with louvers for ventilation.

5.8.1 Monitor lights

These take the form of glass 'boxes' each with a

flat top on flat or low-pitched roofs, the sides are usually arranged to open as with a lantern light.

5.8.2 Thermoplastics

More modern rooflights have been developed using thermoplastics. These include such materials as wire-reinforced PVC, anti-vandal polycarbonate and glass-reinforced polyester resin whilst others offer low-flammability or antiglare qualities. The use of plastic can be very economical and highly versatile with a variety of designs.

5.8.3 Traditional

Traditional rooflights are still framed in metal. Aluminium is mostly used because of its low cost. The glazing in these and almost all rooflights is 6mm wired glass. In the event of a fire, this does reduce the risk of glass falling onto persons below, however, it also delays the venting of a fire as the design ensures that the glass is held together and initially will only crack in fire conditions. To prevent natural heat losses through the roof, some systems employ double-glazing; these systems will also delay the venting of a fire within.

5.8.4 Venting

Some rooflights are intended to form a vent when destroyed by heat i.e., roof ventilators in the form of haystack lantern lights installed above the stage area of theatres. Hazardous conditions can arise however if a rooflight is recessed above a soffit or a suspended ceiling where a fire can develop undetected. Alternatively, heat and flame escaping from a rooflight may reach an adjacent building or flammable materials if it is not well placed. The exposure hazard thus created must be covered in firefighting operations.

It is important to remember however that most rooflights will fail in heat. Smoke, heat and flames can be vented to the outside air thereby ensuring that firefighting can commence in more favourable conditions.

5.9 Stairs and Stairways

Prior to the introduction of building control there was little, if any, control over the construction of stairs and badly designed stairs were fraught with danger in normal use let alone in fire and smoky conditions. The Regulations in England and Wales recognised this and now specify precise minimum dimensions for stairways in dwellings. Also the need to make buildings accessible to disabled people has resulted in ramps and these also come under the control of the Regulations. In Scotland, escape stairs must be constructed of non-combustible materials.

Approved Document B (AD 'B') to the Building Regulations in England and Wales, section 4, covers the design for vertical escape – buildings other than dwellings. An important aspect of means of escape in multi-story buildings is the availability of a sufficient number of adequately sized and protected escape stairs. The relevant building regulations address the number, width and protection of stairs necessary to ensure safe escape. They include measures necessary to protect them in all types of buildings other than dwelling houses, flats and maisonettes

.

In defining the minimum requirement for the safety of persons using a stairway in dwellings etc., the Regulations also differentiate between those for common use – stairways, which serve two or more dwellings, and one dwelling. In all cases stairways and ramps which form part of the structure of the building are required to provide a safe passage for users. In many cases they may represent the only way out of the building in the event of a fire.

The requirements are basically that they should be made from materials of limited combustibility and are continuous leading, ultimately, to a place of safety. This applies whether the stairway is internal or external. A provision is made that combustible material may be added to the upper surfaces of a stairway so that the use of carpets etc., is not precluded.

As a stairway is not designated an element of structure it is not required to have any fire resistance but, as stated above, in most cases must be made of materials of limited combustibility. This leaves a wide range of materials including solid timber and unprotected metalwork.

However, apart from private dwellings and some other residential premises, a good many buildings

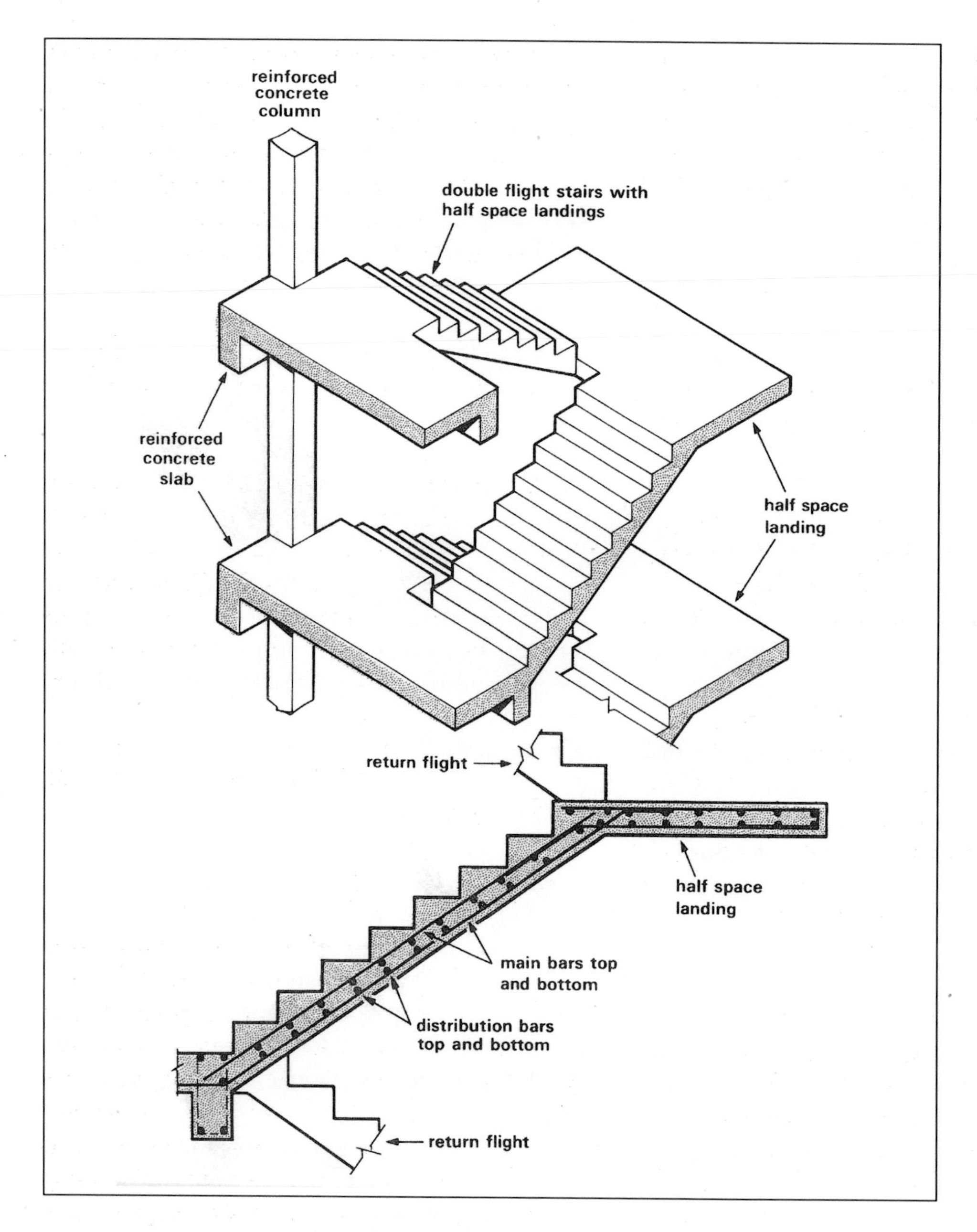

Figure 5.32 Cranked or continuous slab concrete stairs.

where either the public or considerable numbers of people resort, have stairways of stone, concrete or substantial metalwork. An example of a concrete slab stairway is shown in Figure 5.32.

5.10 Trusses

A truss is defined as a framed structure consisting of a triangle or group of triangles arranged in a single plane in such a manner that loads applied at the points of intersections of the members will cause only direct stresses (tension or compression) in the members. Loads applied between these points cause flexural (bending) stresses. The thrust of a truss is downward as compared to an arch where the thrust is outward and usually resisted by a mass of masonry.

5.10.1 The Triangle Principle

The rigidity of the truss rests in the geometric principle that only one triangle can be formed from any three lines. Thus the triangle is inherently stable. An infinite number of quadrilaterals can be formed from four lines, so that the rectangle is inherently unstable.

The economy of a truss is derived from the separation of compressive and tensile stresses so that a minimum of material can be used. There are many designs of trusses.

The top and bottom members of the truss are called chords. The compressive connecting members are called struts. The tensile connecting members are called ties. Connections are called panel points. As a group, the struts, ties and panel points are called the web.

Trusses can be built of wood, wood and steel combined or steel. Cast-iron was used for compression members in early metal trusses or may be found in some 19thC structures. Whilst concrete trusses are not common, they may be found in some very large buildings.

Trusses are most often located within the building. They can also be located above the building to support the roof and ceiling of the top floor so that a large area below can be free of columns.

5.10.2 Problems with Trusses

The truss can be designed and constructed as the minimum structure, which will carry the designed load. Trusses can fail in a variety of ways. All parts and connections of a truss are vital to its stability and the failure of one element of a truss may cause the entire truss to fail.

Multiple truss failures can occur. The failure of one truss can have a serious impact on other parts of the structure, even parts far from the initial failure point. For example steel trusses may be tied together for stability to increase wind resistance. When one truss fails, undersigned stresses by way of the ties may cause multiple failures.

5.11 Walls (loadbearing)

As with steel, walls are usually referred to by the function they perform e.g. external, compartment, separating, and loadbearing. In the following paragraphs some of the types of loadbearing walls are described.

5.11.1 Solid brick

The commonest type of loadbearing wall, and one, which is also widely used as a non-loadbearing panel wall in a framed building, is made of brick. The nominal size of a brick, at least in the south of England, is 228 × 114 × 76mm and the thickness of a brick wall is measured in multiples of a half-brick i.e., 114mm. Thus a "brick-and-a-half" wall is 342mm and a "half-brick" wall is 114mm thick. The bricks are bedded in mortar which may consist of a mixture of lime and sand with water (lime-mortar) or a lime mortar to which has been added a proportion of cement (lime cement mortar or "compo") or of a mortar consisting of cement, sand and water.

Lime mortar is relatively soft and may be protected on the outside of the building by "pointing" the joint with a stronger mortar

Bricks are arranged in a wall so that the vertical joints of one layer, or "course", do not coincide with the joints of the course below. This is known as "bonding" and a number of different arrangements or "bonds" are in general use. The strongest type and the most usual in the UK for thick walls is known as "English Bond". The cavity wall construction is laid in "stretcher bond" with bricks laid lengthways with "snap headers" i.e., bricks cut in half and laid with their ends on the face of the wall to give the necessary bond.

5.11.2 Cavity brick

Cavity walls are used mainly as external walls in buildings particularly exposed to weather. The object of the cavity is to prevent rain penetrating to the inside face of the wall. The usual type of cavity wall found in domestic buildings consists of two half-brick walls held together by metal ties and separated by a 50mm cavity. Sometimes the internal wall is only 76mm thick in modern buildings and is built either of bricks laid on edge or concrete slabs. Whatever method is used the weight of the upper floors, and sometimes the roof, is carried on this internal wall.

The cavity may, or may not, be ventilated to the outside air by airbricks at the top and bottom. In modern building practice the cavity is sometimes filled with an inert material giving additional thermal insulation to the building.

Another type of construction, which uses brick as the "outer skin", is that of modern timber-framed construction. Here the main structure is of timber with the frame clad internally and externally with building board usually of an insulating type. On the outside of this, with a small gap, is laid a conventional brick "skin". A membrane of either plastic or bitumenized paper is placed between the inner and outer skins. In order to prevent fire-spread in the cavity a system of fire stopping is placed at appropriate spots in the cavity. Even if construction is carried out to a good standard, firefighters may have to penetrate the inner skin to locate, and extinguish, a fire in the cavity. This type of construction is not limited to houses but may also be found in two-storey residential homes, hotels, schools etc., Methods of construction vary and firefighters should take any opportunity to inspect these types of buildings because they can present problems.

5.11.3 Timber-framed

It must be stressed that this type of timber-framing is very different to that in the preceding paragraph, and is usually only found on very old buildings. The timbers here are infilled with brickwork, plaster, reeds and plaster (wattle and daub) and various other materials including flints and stones. Again, due to settlement, additions to the building over a long time, alterations internally, rodents etc., any fire can spread through numerous cavities and break out almost anywhere.

5.11.4 Other walls

There are numerous types of solid walls ranging from old stone walls a metre thick to stone-fronted brick walls. Other walls can consist of hollow blocks faced with stone slabs or concrete blocks built up in brick formations and rendered with plaster.

5.12 Behaviour of loadbearing walls in fire

The stability of a brick or stone wall depends, amongst other things on:

(a) Its thickness in relation to its height;
(b) On proper bonding (in particular on the use of sufficient headers to tie the wall together); and
(c) To some extent on its age;
(d) On any horizontal pressure or levering effect which may be exerted on it.

In a stone wall it also depends on the proportion of smaller stones which have been used and the skill of the builder. The fewer the number of joints and the thinner they are then the greater the strength of a stone wall.

A brick or stone wall, though capable of supporting a considerable vertical load, can only withstand a comparatively small sideways or lateral pressure and, for stability, the loading of a wall must be centred within the middle third (Figure 5.33). Provision is usually made in the design of the structure to withstand any normal lateral pressure, either by making the walls themselves thick enough for the purpose, or by the erection of transverse walls or buttresses.

No provision is usually made for abnormal conditions such as may be brought about by fire. The expansion of steel joists may exert lateral pressure upon loadbearing walls into which they are fixed and expansion or other movement of the contents

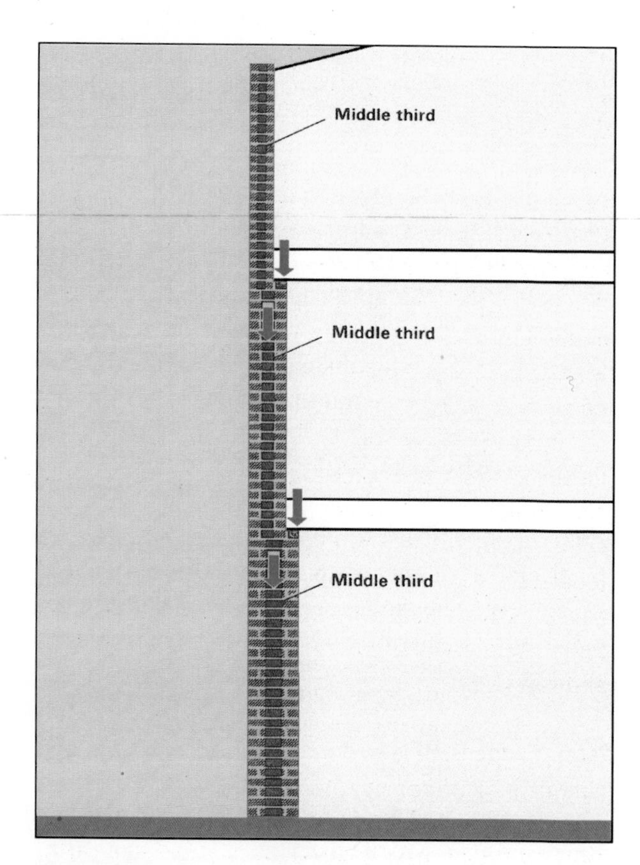

Figure 5.33 Diagram (not to scale) showing how the load on a wall or column must be concentrated within the middle third.

Figure 5.34 Typical light steelwork and fibreboard partition. Photo: Essex Fire Brigade.

of a building may have a similar effect. Both these causes have been known to bring about the collapse of substantial brick walls.

5.13 Non-loadbearing partition walls

The term "partition" or "partition wall" is used when referring to walls whose sole function is the division of a space within a building into separate rooms. In this section the term "partition" is used in the "space division" sense. These partitions are designed and constructed to carry their own weight and any fixture and fittings included in them i.e. doors or glazing. They should also be robust enough mechanically for normal conditions of use, for example, able to resist vibration set up by doors being opened and slammed.

Partition walls are made from a variety of materials. Plasterboard bonded on either side of a strong cellular core to form rigid panels is suitable as non-loadbearing partitions. Compressed strawboard panels provide another alternative method and partially prefabricated partitions such as plasterboard and strawboard can be erected on site.

Non-loadbearing partitions often have quite good sound insulation qualities and will provide a certain period of fire resistance. For example, the fire resistance of timber-framed non-loadbearing partitions will be determined as much as anything by their lining/finishes. Plasterboard linings have an established resistance dependent upon the thickness applied. Plywood and chipboard linings of appropriate thickness also make a contribution to fire resistance.

5.14 Demountable partitions

In certain types of buildings it is often desirable to provide partitioning in a form which can be dismantled and re-erected easily to allow its use in another position. These demountable partitions are usually made up of fairly large components. Steel units or sheet materials such as plasterboard, strawboard and plastic-faced boards are frequently used; connections and angles formed by metal sections are sometimes of extruded light alloy frames (see Figure 5.34).

Firefighters need to be aware that some buildings with this type of partitioning can have their whole internal layout altered without prior notice being given.

5.15 Windows

Windows allow natural light into buildings and also serve to ventilate rooms. Window frames can be made from a variety of materials such as timber, uPVC and metals like aluminium and steel.

Windows can also be designed to operate in various ways by arranging for the sashes (the opening portion of the window including glass and frame) to slide, pivot or swing, by being hung on to one of the frame members. Windows are generally referred to as being of two types according to the method of opening, these being "casement" or "sash".

5.15.1 Casement windows

The simplest consists of a square or rectangular window frame of timber or metal with the window casement hung i.e., hinged at one side. When more than one casement is openable it is usual to refer to them as being two, three or four-light casements. Often a two or more light casement window is in the form of vent lights. These are top-hung to open outwards. The horizontal framing between the casement and the vent light is called the transom. Some windows of this type are constructed so that only part is openable. That part of the window, which does not open, is called a deadlight.

5.15.2 Sash windows

(a) Pivoted sash windows

The opening part of this type of window is supported by pivots at each side, or at the top and bottom, of the frame so that they open partly into, and partly out of, the room. The word sash describes the opening portion of a window and includes the glass and surround.

(b) Sliding sash windows

The vertical sliding sash window with a double hung sash is the most commonly used type of sliding sash window and is constructed so that the two sashes slide vertically in the frame. Another type of sliding sash window consists of a frame in which there are at least two sashes; one or both of which can be opened horizontally.

5.15.3 Double or multiple glazed windows

One sheet of glass in a window is a poor insulator against the transfer of heat. In order to reduce heat loss two sheets of glass, at least 5mm apart, are fixed in the casement or sash with clean dry air trapped between them and often hermetically sealed. Double-glazing does not increase the fire resistance of the glazing to any significant extent and, in fact, this type of glazing can shatter with explosive force when involved in fire. Another arrangement consists of two separate windows, one on the outside opening out and one on the inside opening in. This also serves to reduce heat loss and sound penetration.

In many instances double glazed units contain toughened glass (this is particularly so where large sheets of glass are used). Firefighters should be aware that not only is it virtually impossible for the inhabitants to break this type of window in an attempt to escape but special means need to be employed to break in.

A new glazing concept now used takes full advantage of solar heat. A permanently sealed wafer-thin layer of silver plus other protective coatings give reflective quality with triple glazing efficiency. Other types of glazing are being developed which are not only energy efficient but also minimise air and water penetration and resist the effects of dust, dirt and solvents.

5.15.4 Triple glazing

Triple glazing is an extension of double glazing and is simply the addition of a third sheet of glass giving extra sound and heat insulation.

5.15.5 Leaded windows

Leaded windows consist of a number of small panes of glass held together by strip lead. Such windows, especially those of coloured glass to be found in churches and cathedrals, may be of considerable value. The value lies in the glass not in the leading, which can be replaced. The leaded panes are often held in position by light gauge bars running from side to side of the opening

5.15.6 French windows

French windows are not, strictly speaking, windows but should be regarded either as a pair of panel doors or as casement windows.

5.16 Doors, Doorsets and Shutters

Doors and shutters are of seven principal types:

- Hinged doors
- Swing doors
- Revolving doors
- Sliding doors

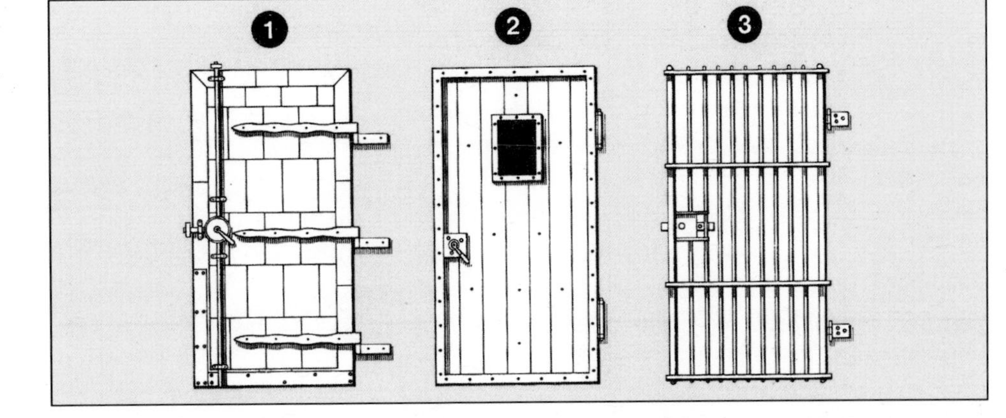

Figure 5.35 Typical metal doors: (1) and (2). Two types of steel-covered fire-resisting door.
(3) Barred door.

- Folding doors
- Cantilever doors
- Roller shutters.

In many gates and doors of industrial premises, a small door, often referred to as a 'wicket door', may be inset, e.g. a hinged door set in a sliding gate.

5.16.1 Hinged doors

Hinged doors closing against a rebate on the door-jamb are by far the most common. Types of hinged door are:

(a) Flush

Probably the commonest type of hinged door, and one, which is relatively cheap to construct, is the lightweight flush door. This usually consists of two layers of plywood or hardboard with a honeycomb paper core. Sometimes the core is merely strips of strawboard glued on. A number of cross members may strengthen the hollow door; alternatively, some better types of flush door are solid.

(b) Panelled

Panelled doors usually have a wooden frame with wooden, or sometimes in the upper half, glass panels. There may be, in all, four panels, two small and two large.

(c) Ledged

Many ledged doors are of light construction. They may be ledged only or there may be bracing in addition or framing – a common type is framed, ledged and braced.

(d) Metal
Examples of steel-covered doors are shown in (Figure 5.35 (1 & 2). Doors of this kind may sometimes be of steel with wooden linings, so that the steel is concealed. Barred doors vary greatly in construction, but a typical example is shown in Figure 5.35 (3).

5.16.2 Swing doors

Any of the above types of door may be found with special pin hinges allowing them to swing in either direction, and consequently there is no rebate on the jamb. Such doors may consist of a single or a double leaf in a single opening. Swing doors are frequently used in restaurants, hotels and department stores, and in long corridors to check the spread of smoke in case of fire. They are often partly glazed, the glass being wired or set in copper glazing bars or intumescent paste in those instances where some degree of fire resistance is required. If solid, they frequently have a glass panel and are generally of flush construction. In large department stores and modern office blocks, frameless swing doors of toughened glass may be encountered.

5.16.3 Revolving doors

Revolving doors present an obstruction to the fire-fighter since, unless they are first collapsed or broken in, they do not permit the passage of bulky objects or lines of hose. Revolving doors turn on central pivots at the top and bottom and usually have four wings arranged at right angles to one another. In some types of door only two wings may be found, each of which has a curved extension piece. The wings on the doors are generally constructed to collapse and to move to one side so as to give a relatively unobstructed opening. It is essential that this type of door be used in conjunction with a normal hinged door.

There are two common methods of securing the wings of revolving doors. In the first, the wings are usually held in place by a bracket or solid stretcher bar situated usually at the top of the door across the angle formed by the leaves where they join the newel post. One end of each bar is permanently connected to one leaf of the door and the other engages with some form of catch on the adjacent leaf. The wings are collapsed by releasing the stretcher bars, either by undoing the wing nuts or by unfastening the catches, which hold them in place.

In the second type, the two opposing wings are hinged to the single leaf formed by the other two and are kept in position by a chain which runs through them and is held by a catch on each of the hinged leaves. If this catch is released, the wings can be folded back to give a clear opening.

5.16.4 Sliding doors

These doors may be either of solid construction or in the form of a lattice, which collapses into a relatively small space when opened. Sliding doors may run on tracks above and below the door or be suspended from an overhead track. They are not often encountered in domestic property except possibly in garages, but those of solid construction are widely used in commercial premises, especially as fire-resisting doors for isolating sections of a building. These doors may either slide on one or both sides of the opening, or alternatively may move into a central recess in the wall.

Steel lattice doors are widely used to protect property where weatherproofing is unimportant. The gate usually runs on two sets of tracks, one above and the other beneath, but may sometimes be found with a bottom track only. They are often to be found as a protection to the opening of a lift or lift shaft.

5.16.5 Folding doors

Folding doors are usually of fairly light construction, but exceptionally, they may be very large and of robust construction. They are often found as separating doors between two rooms where space is valuable. They are similar in design to normal hinged doors, but the two or more leaves are hinged together so that the whole door opens to one side only.

5.16.6 Cantilever doors

This door is counterbalanced and pivoted so that the whole doors rises upwards and, when open lies horizontally. Cantilever doors are usually found on garages, but steelplated doors of this type are also found in boiler houses. These doors generally fit flush in the opening.

5.16.7 Roller shutters

Roller shutters are nearly always made of steel, but may be constructed of timber. Small roller shutters can be raised by hand, but the larger sizes are almost invariably operated by means of gearing and some form of handle or chain and block on the inside.

5.17 Fire Resistance of Doorsets

A doorset is defined as an assembly (including door and frame or guide) for the closing of permanent openings in separating elements. For the purposes of British Standards, the term doorset includes shutter assemblies, but excludes fire dampers for incorporation into ducts.

In England and Wales, Approved Document 'B' – Fire Safety, to The Building Regulations 1991 gives the specification of fire doors. The doors are identified by their performance under test to British Standards, in terms of integrity for a period of minutes, e.g. FD30 (Fire Door 30 minutes). A suffix (S) is added for doors where restricted smoke leakage at ambient temperatures is needed.

The minimum fire resistance of doors in terms of integrity is given in Table B1 – Provision of fire doors of appendix B of Approved Document 'B'. The location of the door determines the minimum fire resistance and this includes:

(a) In a compartment wall separating buildings;
(b) In a compartment wall;
(c) In a compartment floor;
(d) Forming part of the enclosure of a protected stair or lift shaft;
(e) Forming part of the enclosure of a protected lobby or protected corridor to a stairway, any other protected corridor or a protected lobby approach to a lift shaft;

(f) Sub-dividing corridors connecting alternative exits or dead-end portions of corridors from the remainder of the corridor;
(g) Any door within a cavity barrier, between a dwelling house and a garage or forming part of the enclosure to a communal area in sheltered housing.

The method of test exposure is from each side of the door separately, except in the case of lift doors which are tested from the landing side only.

All fire doors should be fitted with an automatic self-closing device except for fire doors to cupboards and to service ducts, which are normally kept locked shut.

Where a self-closing device would be considered a hindrance to the normal approved use of the building, self-closing devices may be held open by:

- A fusible link, but not if the door is fitted in an opening provided as a means of escape (unless it complies with (d) below);
- An automatic release mechanism actuated by an automatic fire detection and alarm system;
- A door closure delay device; however,
- Two fire doors may be fitted in the same opening so that the total fire resistance is the sum of their individual fire resistances, provided that each door is capable of closing the opening. In such a case, if the opening is provided as a means of escape, both doors should be self-closing, but one of them may be fitted with an automatic self-closing device and be held open by a fusible link if the other door is capable of being easily opened by hand and has at least 30 minutes fire resistance.

Unless shown to be satisfactory when tested as part of a fire door assembly any hinge on which a fire door is hung should be made entirely from materials having a melting point of at least 800°C.

Other than fire doors:

(a) Maisonettes;
(b) Bedroom doors in 'other residential premises' (defined as hotel, boarding house, residential college, hall of residence, hostel or any other residential purpose not described in purpose group 2(a) Institutional);
(c) Lift entrance doors.

All fire doors should be marked with the appropriate fire safety sign complying with British Standards according to whether the door is:

(i) To be kept closed when not in use;
(ii) To be kept locked when not in use; or
(iii) Held open by an automatic release mechanism.

Fire doors to cupboards and to service ducts should be marked on the outside. All other fire doors should be marked on both sides.

Apart from its normal function, a standard door will:

(a) Serve to contain an outbreak of fire; and
(b) Will prevent the penetration of toxic smoke and fumes into otherwise unaffected parts of the building for a short period of time.

A closed door also restricts the flow of oxygen thereby helping to starve the fire. It is for these reasons that all doors should be kept shut particularly when a building is unoccupied for any length of time or at night. To be given a fire resistance rating, a complete doorset must be specially designed and built.

5.18 Function of fire doors

A fire door may be defined as:

A door or shutter provided for the passage of persons, air or things which, together with its frame and furniture as installed in a building, is intended, when closed, to resist the passage of fire and/or gaseous products of combustion and is capable of meeting specified performance criteria to those ends.

Fire doors have at least two functions:

(a) To protect escape routes from the effects of fire so that occupants can safely reach a final exit: and/or
(b) To protect the contents and/or the structure of a building by limiting the spread of fire.

Consequently a particular fire door may have to perform one or both of these functions for the purposes of smoke control, protecting means of escape, compartmentation or the segregation of special risk areas.

Fire doors provided for smoke control purposes are designed to restrict smoke movement and should be capable of withstanding:

- Smoke at ambient temperatures; and
- Limited smoke at medium temperatures.

Smoke control fire doors are provided for life safety purposes and play an important role in the vicinity of the fire in its early stages and in protecting escape routes more remote from a fully developed fire. There is, at present, no criteria for smoke control doors although a recommendation for performance for fire doors to resist the passage of smoke is under consideration.

Fire doors provided to protect means of escape should:

- Be capable of achieving a minimum fire resistance for integrity of only 20 minutes;
- Withstand smoke at ambient temperatures;
- Withstand limited smoke at medium temperatures.

These types of doors are required to keep escape routes sufficiently free from smoke for a sufficient time for occupants to reach a place of safety and to maintain integrity against the effects of fire for long enough to fulfil that objective. It follows that, in addition to smoke control capability, these doors require a measure of fire resistance. Consequently Codes of Practice generally recommend such doors to have either 20 or 30 minutes fire resistance and to have both flexible edge seals and heat activated seals i.e. intumescent strips.

Doors for compartmentation and segregation of special risks are doors which must be capable of achieving the period of fire resistance appropriate to the structure which is not less than 30 minutes and may be as much as 4 or even 6 hours in exceptional circumstances. The required fire resistance may need to be achieved by the provision of two fire doors in series, both having half the required fire resistance. If such doors are required to protect an escape route they will need also to have the smoke control capabilities described above.

5.19 Identification of fire doors

Fire doors should be identified by the initials FD followed by the performance in minutes that the door should achieve when tested for integrity only. For example a door identified as FD30 implies integrity of not less than 30 minutes i.e. 30 minutes fire resistance.

Where the door should also resist the passage of smoke at ambient temperatures the suffix 'S' should be added to the identification.

The practical application of the above is, for example, when specifying for a flat entrance door in a block of flats where the door would be required to protect a means of escape (see above) the requirement would be for an FD30S door. Or, alternatively, a fire door required in a compartment wall of 60 minutes fire resistance, with no means of escape implication, would have to be an FD60 door.

5.20 General

The fit of a door in its frame is a significant factor. The doorstop of a 60-minute door is required to be cut from the solid timber frame whereas, in a 30-minute door, it is permissible for the stop to be screwed onto the frame. The weak point of a door in a fire is often the face on which the hinges are exposed and particularly the hinge side. It is important that hinges are made of non-combustible material and they are required to have a high melting point (800°C). Where unlatched doors are used for smoke control purposes the selection of the self-closing device is critical.

When other methods of smoke control are provided e.g., pressurisation, the smoke control criteria for doors may not be applicable. It must not, however, be assumed that all doors must be fire resisting. For example, it would be futile to require a door to be of a fire-resisting standard where the partition in which it is fitted is not required to perform any smoke or fire retardant function but is simply a convenient subdivision of space.

Chapter 6 – Building Materials – Other elements of structure

6.1 Asbestos

Asbestos has been used in building materials for many years and there is a large tonnage of asbestos materials in existing buildings. The use of asbestos in new building materials has been sharply reduced in recent years, but a significant tonnage is still used, mainly in the manufacture of asbestos-cement.

Asbestos is a fibrous material which occurs in many parts of the world, it is estimated that a total of about 6 million tonnes have been imported into the United Kingdom over the last century.

The three main types of asbestos produced commercially are:

(a) Crocidolite – blue asbestos
(b) Amosite – brown asbestos
(c) Chrysotile – white asbestos.

The import of crocidolite into the UK ceased in 1970 and the import of amosite in 1983.

In addition to the general duty placed on employers by Section 2(d) of the Health and Safety at Work etc. Act 1974 (Chapter 37), there have been several sets of Regulations made under the 1974 Act and the Consumer Safety Act 1978 (Chapter 38) to implement several of the requirements of two European Community Directives (Council Directive 83/477/EEC and Council Directive 83/478/EEC).

In addition, certain provisions of the Public Health Acts are relevant to the use of asbestos materials in the construction and demolition of buildings. The Public Health Act 1936 (Chapter 40) Section 92 et seq, as amended by the Local Government (Miscellaneous Provisions) Act 1982 (Chapter 30) defines a number of statutory nuisances. Section 29 of the Public Health Act 1961 (Chapter 64) as amended by the Local Government (Miscellaneous Provisions) Act 1982 gives local authorities powers to impose conditions on the demolition of buildings. In Scotland, similar controls are exercised under the Building (Scotland) Act 1959.

The following are examples of ways in which asbestos materials are, or have been used in buildings. The installation of sprayed asbestos and thermal and acoustic insulation is now prohibited and asbestos insulation board is no longer manufactured in the United Kingdom. However, these materials may still be present in some buildings. Asbestos is still used in the manufacture of asbestos-cement and in materials such as mastics, sealants, roofing felts and protected metals.

6.1.1 Sprayed coatings and lagging

The sprayed material applied in the UK was a mixture of hydrated asbestos-cement containing up to 85% asbestos fibre. The application of sprayed asbestos ceased in 1974 and is now prohibited by the Asbestos (Prohibitions) Regulations.

Lagging is a term which covers a wide range of materials including pipe sections, slabs, rope, tape, corrugated asbestos paper, quilts, felts, blankets and plastered cement.

Asbestos has been used as a surface coating on felt and cork insulation. Asbestos lagging may have a protective covering of cloth, tape, paper or metal. The installation of asbestos thermal insulation is now prohibited by the Asbestos (Prohibitions) Regulations which came in to effect on 1 January 1986.

6.1.2 Insulating Boards

Insulating boards have a density of approximately 700 kg/m3 and contain 16-40% asbestos mixed with hydrated Portland cement or calcium silicate. Asbestos insulating board ceiling tiles were made of board cut into squares and were widely used in schools, hospitals and shops.

6.1.3 Ropes, yarns and cloth

The asbestos content of woven and spun materials approaches 100%. Asbestos yarns, often reinforced with other yarns or filaments, were used in jointing and packing materials, gaskets and caulking for brickwork. Asbestos ropes have been widely used for thermal insulation of pipes particularly where pipes pass through fire resisting walls.

6.1.4 Millboard, paper and paper products

These materials have an asbestos content approaching 100% and have been used for insulation of electrical equipment and for thermal insulation.

6.1.5 Asbestos-cement products

Asbestos-cement products generally contain 10-15% asbestos fibre bound in a matrix of Portland cement or autoclaved calcium silicate. Corrugated sheet is largely used as roofing and cladding. Semi-compressed and fully compressed flat sheets are largely used as panelling or partitions, the degree of compression largely determining the strength of the material.

The asbestos fibres in asbestos-cement are firmly bound in the cement matrix and will be released only if the material is mechanically damaged or deteriorates with age.

6.1.6 Bitumen felts and coated metals

Some roofing felts, flashing tapes, damp-proof courses and other products contain asbestos fibre, sometimes in the form of asbestos paper, in a bitumen matrix.

6.1.7 Flooring materials

Asbestos has been added to the mix of certain PVC and thermoplastic floor tiles and sheet materials. Some types of PVC flooring have an asbestos paper backing

6.1.8 Textured coatings and paints

Asbestos may be found in some textured coatings or paints available for both interior and exterior decoration.

6.1.9 Mastics, sealants, putties and adhesives

Small quantities of asbestos may be included in mastics, weatherproofing sealants, putties and adhesives to impart anti-slumping characteristics, to improve covering power and prevent cracking or crazing.

6.1.10 Reinforced plastics

Asbestos-reinforced PVC has been used to make cladding and panels. Asbestos-reinforced plastics have also been used to make a variety of products, including household items such as plastic handles and battery cases.

Asbestos has had a variety of uses in the home, in schools and colleges. In addition to building construction it has been used in some domestic appliances, household goods and materials for DIY work. It may have been used in fixtures and fitting such as fireplace surrounds, the lagging of older central heating systems, asbestos-cement flue pipes and asbestos millboard covers for fuse boxes.

The mere presence of asbestos materials does not constitute a hazard, and removing undamaged material may release more dust than would leaving it in place.

6.1.11 Domestic appliances

Asbestos has been used in domestic appliances for its thermal and electrical insulating properties. UK manufacturers no longer use asbestos in hair driers, fan heaters, irons, toasters, washing machines, tumble driers, spin driers, dish washers, refrigerators or freezers, except for small amounts in some appliances in the form of gaskets or braking pads, which are usually sealed within the appliance and are unlikely to release free fibres into the atmosphere.

Asbestos may still be found in some older types of cooker as oven linings (asbestos board), sealing between metal plates in the oven (asbestos fire cement) or oven door seals (asbestos rope).

The use of asbestos is being phased out by the manufacturers, although asbestos rope is still available and may be used by some service engineers.

6.1.12 Household goods

Asbestos textiles are no longer used in oven gloves manufactured in the UK. Simmering mats made from asbestos millboard or asbestos paper may still be available. Older types of ironing board may have had rests made from asbestos millboard or paper but these have largely been replaced by asbestos cement.

6.1.13 Asbestos fire blankets

It is quite possible that asbestos fire blankets, once commonly found in laboratories, schools and catering establishments, may still be in use. HSE has issued advice on dealing with any such blankets which should be followed to avoid fibre release.

6.1.14 Asbestos in heating systems

Asbestos has been used for thermal and electrical insulation in a number of different types of heating systems including catalytic gas heaters, electric storage heaters and gas warm air heaters. Various types of asbestos have been used at different times and some models contain asbestos and others not.

Some types of gas heaters, fuelled by liquefied petroleum gas, contain loosely compressed asbestos fibre panels. These catalytic gas heaters 'burn without a flame' – the gas is oxidised as it passes over the asbestos panel, which contains a platinum catalyst. The panels may contain up to 0.5kg of chrysotile asbestos.

Asbestos insulating board and other asbestos based materials have sometimes been used for fire protection in the construction of cupboards under staircases and heater cupboard doors

6.1.15 Electric 'warm air' and storage heaters

Asbestos materials have been used to provide insulation in older models of electric storage heaters and larger warm-air heating systems. In the main three types of appliances were involved:

(a) Storage heaters which operate primarily by radiant heat;
(b) Fan-assisted storage heaters which operate by radiant heat and convection;
(c) Through the heater by a fan and distributed directly or via ducts.

6.1.16 Do-it-yourself work

DIY applications for asbestos span a wide range of products, the asbestos content ranges from 0.5–2% in putties to over 90% for some wall plugging compounds. Common applications include:

- Wall plugging compounds;
- Textured paints;
- Flashing paint;
- Bitumen sealants;
- Roofing felt;
- Damp-proof courses;
- Mastics;
- Floor tile adhesive;
- Vinyl floor tiles;
- Cushion flooring;
- Asbestos rope or string.

Once in place and in normal use, these products present very little hazard in the home.

6.1.17 Asbestos in fires

Asbestos fibres change their mineral structure after prolonged heating at high temperatures, losing their fibrous nature and mechanical strength, and becoming less hazardous. However, little is known about the fate of asbestos building materials under actual conditions of a severe fire.

There is evidence that debris containing unchanged asbestos fibres can be dispersed over a wide area and can travel considerable distances from the seat of a major fire. Debris containing unchanged asbestos fibres may also remain at the seat of the fire itself. Severe fires and explosions in buildings which are clad or roofed with asbestos-bitumen coated metal may cause the coating to burn off from the metal, generating a grey paper-like ash which may contain unchanged asbestos fibres.

6.2 Building Boards

Whatever the cause of a fire within a building, the surfaces of wall and ceilings will contribute to the fire if they are ignited. Ensuring that these surfaces are non-combustible or at least, very difficult to ignite, means that, if a fire does start, its rate of growth will be reduced. It is also desirable to keep to a minimum the amount of smoke or toxic fumes given off from this type of surface if involved in a fire, particularly those lining escape routes.

Wall and ceiling linings are normally classified in guidance by reference to two British Standard test methods. These are the spread of flame test and the fire propagation test. The spread of flame test has four classes, these are class 1 to class 4, with 1 being the highest performance rating. Class 0 is a further class that is used for critical situations where a higher standard of performance than that of class 1 is appropriate. It is not a British Standard test classification (although it is referred to in British Standards on fire precautions in buildings).

A wide range of materials are used in the manufacture of sheet material used for this end-use application and they are of varying sizes, thickness and fire resistance. They bear many different trade names but may be classified generally in one of the following groups:

(a) Fibre building boards
(b) Plaster boards
(c) Asbestos boards
(d) Plywood boards
(e) Block boards
(f) Plastic boards.

6.2.1 Fibre building boards

Fibre building boards are manufactured in a wide range of sheet materials, usually more than 1.5mm thick. They are made from actual wood fibres or woody plants and derive their basic strength and cohesion by the felting together of the fibres themselves, and from their inherent adhesive properties. Bonding, impregnating or other agents, including fire retardants, may be added during or after manufacture to modify particular properties.

Fibre building boards fall into two major groups according to whether the board has been compressed in a hydraulic press during manufacture or not. The non-compressed type is termed insulating board (softboard). This is used in sheet form and as tiles. Bitumen impregnated insulating board also comes within this category; it is used for sheathing timber-framed buildings and for roof-sarking (lining), the bitumen content gives it a high resistance to moisture.

In the second group are medium boards of low or high density from 6–13mm thick and hardboards. Standard hardboard is a dense sheet material 2–13mm thick with one smooth face and a mesh pattern on the reverse. Tempered hardboard, 3–13mm thick has high strength and water resistance. It is made by impregnating standard hardboard with oils and resins, usually immediately after pressing, and then applying further heat treatment. Building boards of this group are not easily ignitable but all are combustible.

6.2.2 Plaster boards

Plaster boards for interior use are composed of a core of set gypsum or anhydrite plaster enclosed between, and firmly bonded to, two sheets of heavy paper to increase their tensile strength. In a fire the exposed paper face may burn away making it relatively easy to break up the non-combustible gypsum core, but until this happens, the plaster board will retard the spread of fire.

Vermiculite is a clay-mineral which expands to many times its original volume when subjected to high temperature. It is incorporated in plasterboards to give it a superior fire-resistant rating than ordinary linings. Plasterboard and vermiculite can also be mixed with other products such as silicate which again is non-combustible and does not emit smoke when involved in fire.

6.2.3 Asbestos boards

It goes without saying that asbestos does not form a part of the composition of today's products but cement sheets or insulating or wallboards may still be found in older buildings. When subjected to fire, the amount of smoke given off is negligible.

6.2.4 Plywood boards

Plywood boards are made up of thin wood laminations laid in alternate directions to increase their strength. Their susceptibility to fire depends on the type of timber used and the overall thickness of the board. The type of bonding material may have some bearing on the development of a fire.

6.2.5 Block boards

These are made from a core of separate wood blocks bonded together and finished externally with a veneer or plastic overlay to give the appearance of a homogenous board. They are produced in many grades and qualities and their behaviour in fire varies accordingly.

6.2.6 Plastic boards

Plastic boards are composed of organic materials, e.g. paper, linen, sawdust or woodchips, bonded together with synthetic resins and subjected to heat and pressure. Phenolic laminates are rigid boards made of sheets of special paper impregnated with phenol-formaldehyde and urea-formaldehyde. This type of board has good fire-resisting properties and usually incorporates a flame retardant substance in its manufacture. Resin-bonded sawdust (or woodchip) boards are sawdust and/or woodchips bonded with synthetic resins, and are man-made timbers; their behaviour in fire is dependent on their surface treatment.

There are many other types of popular plastics available. One which has increased in use is expanded polystyrene which is often used as wall and ceiling tiles because of its good thermal insulation qualities. Although this is available in flame retardant grades it is known to burn fairly rapidly and often softens and collapses. Foamed polyurethane in flexible and rigid forms are also excellent thermal insulators and are used in various applications as a weather-resistant coating. Again this is available in flame retardant grades but generally burns rapidly producing thick dark smoke.

6.3 Building blocks and Slabs

6.3.1 Building blocks

Building blocks, like bricks, are used for the construction of walls of all types, and they have become popular because of the savings resulting from the improved productivity when laying units larger than bricks.

Blocks are generally made of concrete combined with various types of aggregates which give the block different loadbearing qualities whilst others are designed purely for insulating qualities.

Hollow-fired clay blocks combine a clay aggregate to produce a particularly lightweight block. The hollow interior is filled with polyurethane foam to give it excellent thermal properties.

Their fire-resisting qualities are generally better the greater the thickness and the smaller the proportion of voids. In a fire, the face of the block exposed to the fire, whether used in a partition or a floor, may spall as a result of the unequal expansion of the material in the block as the temperature rises.

There are several types of concrete block which are made in a variety of thickness from 50mm to over 100mm thick. Their size and whether they are solid or hollow decide if they are to be used as loadbearing walls or non-loadbearing partitions. Most are moulded by special machines from concrete made with normal dense or lightweight aggregates.

Aerated concrete is made by a completely different process using cement and sand and/or pulverised-fuel ash (PFA) or cement and lime. The addition of fine aluminium powder causes the formation of numerous small air cells. A large 'cake' is produced which is cut into pieces and autoclaved (high pressure steam curing).

For the purpose of determining fire resistance, machine-made blocks are divided into two classes according to the type of aggregate used in their manufacture:

(a) Class 1 blocks – those with a higher fire resistance for a given thickness – are made from lightweight aggregates;
(b) Class 2 blocks which, for the same period of fire resistance, require a slightly greater thickness, are made from naturally dense aggregates other than limestone.

Slightly different values apply to aerated concrete blocks; compared with Class 1 aggregate blocks, loadbearing walls with 240 or 360 minutes of fire resistance should be a little thicker, but non-loadbearing walls with low periods of resistance can be thinner. All types provide a high degree of fire resistance with little risk of collapse or deterioration and, therefore, give effective compartmentation (see Figure 6.1).

Figure 6.1 Blockwork wall severely distorted but still intact following a fire. Photo: Cheshire Fire Brigade.

The fire resistance of block walls is improved if they are plastered on both sides and especially so if a lightweight plaster, such as vermiculite-gypsum plaster, is used.

6.3.2 Building slabs

Building slabs can come in a variety of sizes and are generally made out of long wood fibres mixed with Portland cement and compressed. They are combustible but the wood is chemically treated to provide fire resistance and often are water resistant as well. Slabs are used for roof decking and heat insulation.

6.4 Cement (including Glass Reinforced Cement (GRC))

Cement is a fine powder – usually made with various types of Portland cement – which forms part of a combination of materials to make concrete. It reacts chemically with water and the longer the drying process the more strength is developed.

Glass reinforced cement, with its rapidly increasing applications in construction, is a composite material consisting of cement and a small proportion of glass fibres. The addition of GRC reinforcement enhances the strength and toughness of the cement; however, there is some doubt as to the long term durability of GRC.

Two particular characteristics of cement can be changed by the addition of small proportions of other materials. For example, cement is inherently fire resistant but the addition of pulverised fuel ash (PFA) can increase that fire resistance. Cement is also known to shrink considerably when drying and a small amount of sand added to the GRC ensures that this does not happen.

6.5 Concrete

Concrete is a cementious material produced by a chemical reaction of Portland cement and water to which inert materials called aggregates such as sand, gravel or crushed stone, are added.

A common error is the use of the term 'cement' when 'concrete' is the proper term, cement is a component of concrete.

Shortly after it is mixed, concrete sets into a solid, rock like mass, but it continues to cure indefinitely. Construction specifications usually set a date by which concrete must reach its required compressive strength. For instance, concrete required to reach design strength in 28 days may be referred to as '28 day concrete'. Other concrete mixes may achieve full compressive strength in less time.

Whilst curing, concrete generates heat of hydration. During its initial curing, concrete must be protected from freezing as low temperatures retard the curing and freezing is harmful to the material.

Concrete is very weak in tensile strength and has poor shear resistance. Its compressive strength is good, particularly when compared to the cost of steel to resist the same load.

Concrete is inherently non-combustible, it may have been fabricated to meet a specific fire-resistance standard. Unfortunately there is often confusion over non-combustibility with fire-resistance. Neither is synonymous with fire safety.

Concrete can be produced having a wide range of properties such as high compressive strength, durability, thermal insulation and fire protection. All these qualities largely depend on the materials and the proportions used in the mix. When concrete is heated, it expands due to thermal expansion of the materials, but the hardened cement paste also shrinks as a result of loss of moisture by drying out. As a result the overall change is not easily predicted and internal stresses can be set up within the concrete.

In a severe fire, spalling of the surface material occurs and is aggravated if the hot concrete is suddenly chilled, for example with a jet of water. Concrete made with limestones or lightweight aggregates, are very much less susceptible to spalling than those made with more dense aggregates, hence the fire resistance of structural concrete is classified differently according to the type of aggregate used.

It is possible to achieve very high levels of fire resistance with reinforced concrete, up to four hours is easily achieved. However, as reinforced concrete depends for its tensile strength on the steel reinforcement, it is critical that in the design of the elements sufficient protection is provided to the steelwork. Simply increasing the thickness of the concrete cover to the reinforcement does not necessarily give the corresponding increase in safety because of the tendency of concrete to spall (break off) in a fire. This can reduce the cover and it may be necessary to provide supplementary reinforcement to counteract this danger if the cover is thicker than 40mm.

One of the critical issues in the fire resistance of concrete is the nature of the aggregate which is being used, certain aggregates being more resistant to spalling and having a lower thermal conductivity. The issue of thermal conductivity is particularly important when the assembly is also providing a subdivision and it is necessary to limit heat transfer. Also critical can be the use of permanent steel shuttering when it is necessary to design concrete slabs to be able to withstand the failure of the steel.

There are two basic types of concrete construction: cast-in-place which includes plain concrete, reinforced concrete and post-tensioned concrete; and precast which includes plain concrete, reinforced concrete and pre-tensioned concrete.

Within the types of concrete construction, certain definitions are used, they include:

(a) Aggregate – this material is mixed with cement to make the concrete. Common aggregates are both fine and coarse. Fine aggregate is usually sand. Coarse aggregate may depend on the desired characteristics of the finished product and include crushed stone, gravel, shale, slate or clay.
(b) Cast-in-place concrete – this concrete is moulded in the location in which it is expected to remain.
(c) Casting – the process of placing fluid concrete into moulds, generally called forms, in which the concrete is permitted to harden to a certain shape.
(d) Plain concrete – this term refers to concrete that has no reinforcement.
(e) Pre-tensioning and post-tensioning – these are processes by which steel rods or tendons are placed under tension drawing the anchors together. The tensioned steel places the concrete in compression.
(f) Precast concrete – this concrete has been cast at a location other than the place where it is to remain.

(g) Pre-stressed concrete – pre-stressing places engineered stresses in architectural and structural concrete to offset the stresses which occur in the concrete when it is placed under load.
(h) Reinforced concrete – this is a composite material made of steel and concrete. Steel provides the tensile strength that concrete lacks.
(i) Spalling – this term describes the loss of surface material when concrete (or stone) is subjected to heat. Some concrete and certain aggregates are more subject to spalling than others.

Concrete may be in one of the following forms:

6.5.1 Reinforced concrete

Except for concrete bricks and blocks, concrete is rarely used for structural purposes without being reinforced because it is relatively weak in tension and prone to crack. In the early days of reinforced concrete construction, reinforcement consisted of plain round mild steel-bars, but high tensile steel reinforcement – hot rolled bars or cold-worked bars of different types – have been introduced over the years. In reinforced concrete, the steel is not stressed until loads are imposed on the structural member.

6.5.2 Pre-stressed concrete

Pre-stressed concrete is a form of structural concrete in which tensile steel tendons are stressed against the length of concrete, which is thus put into compression, before imposed loads are applied. Pre-stressed concrete is sub-divided into pre-tensioned and post-tensioned systems.

(a) Pre-tensioned concrete
This has the tendons stretched and anchored independently of the concrete before the concrete is cast around them and allowed to harden. The tendons are then released from their anchorage but, because they are now bonded to the hardened concrete, they are anchored by the concrete and put it into compression.

(b) Post-tensioned concrete
This is cast with ducts through which the tendons are threaded and then stressed after it has hardened, each tendon being anchored against the concrete. The tendons may remain unbonded, but often the space between the tendons and the ducts is grouted so that the tendons become effectively bonded to the concrete and at the same time are protected against corrosion.

No distinction is drawn between the different forms of pre-stressed concrete in assessing their fire resistance.

6.5.3 Fire resistance of concrete

As was mentioned earlier, the fire resistance of concrete itself is determined by the aggregate used in its make-up. However, the fire resistance of structural concrete, whether reinforced or pre-stressed, is determined primarily by the protection of the steel against an excessive rise in the temperature. This is afforded by the concrete cover, i.e. the concrete between the surface of the member and the nearest surface of the embedded steel. Generally, the greater the amount of cover, the longer the period of fire resistance.

The so-called 'critical' temperature for steel is about 550° Celsius for mild steel and about 400°C for high tensile steel, but in neither case is there a sudden change in the properties of the steel. These are the temperatures at which they lose about half their cold strength and therefore at which most of the design factor of safety is likely to be used up. In a fire, structural concrete does not normally collapse suddenly – it may deflect considerably under load, and floors may suffer local break down, but even after a severe fire most concrete structures are safe enough to be reinstated to perform their original functions

6.6 Glass

Glass is non-combustible and will not, therefore, contribute fuel to a fire or directly assist a fire to spread. At one time, glass would have constituted a major weakness in a wall, door or screen because it would break and fall out. Fire resisting glazing not only provides an effective barrier to the spread of smoke and flames, it can also provide protection against radiant heat in a fire.

Architects and designers are no longer restricted by solid materials such as timber and metals. Today in its various fire safety forms, glass offers new dimensions and designers can create the feeling of more light and space with glass. In high security areas such as Banks and Building Societies, the glass may be specially toughened to resist vandal attacks.

Until recently, limitations were placed on the use of glass as an element of building construction in fire resisting applications. This was due to the natural property of glass to directly transmit radiation. When a fire is fully developed, high levels of transmitted radiation through glass may present a hazard to people escaping or possibly cause the ignition of combustible materials resulting in fire spread. Thus, a window in a compartment wall could be a weak link.

However, clear vision through a wall or door can also provide safety benefits in the event of a fire, enabling the location and safe evacuation of occupants from burning buildings. This is particularly true in fire doors and has led to the extensive use of wired glass in these locations.

Normal glass has very little fire resistance, offering little insulation and being liable to lose its integrity and stability as it shatters under fire conditions. However, there are three types of glass now available which offer some degree of fire resistance.

The familiar Georgian-wired glass can solve the problem of stability and integrity by holding the glass in place, but this still does not offer any insulation and radiant heat can still pass through the material. Wired glass permits the transmission of heat radiation from the fire, yet provides specific resistance to the passage of flames and smoke; it is a non-insulating glass. Other non-insulating, non-wired glass products with similar properties have also become available in recent years owing to improvements in glass processing technology. Toughened glasses are now available (e.g. 'Pyran' from Schott and 'Pyroswiss' from Colbrand) which achieve the same integrity and stability as the wired glass without the unattractive appearance of the wires, yet these also fail to provide any insulation.

A second, more recently developed range of glass products is able to offer enhanced fire performance characteristics.

The one type of glass which does offer insulating properties is laminated glass, e.g. 'Pyrostop' from Pilkington and 'Pyrobel' from Glaverbel. These incorporate a completely translucent and transparent intumescent layer, which, on the application of heat expands to form an insulating barrier. These glass products have the ability therefore, to resist fire radiation transmission, by becoming opaque when subjected to heat above 120° Celsius. Such fire insulating glass products enable glass to be used in larger sizes and in locations which were previously served by traditional opaque fire compartment materials, such as brick or concrete.

The disadvantages of such laminated glass lie in its weight, cost and limitations on external use. Such glass must be ordered pre-cut, and this is a factory, rather than a site job. With all three types of glass (wired, toughened and laminated), the design of the frame is as important as the choice of glazing material and it is essential that the frame will survive as long as the glass. It is crucial that the architect considers the fire resistance of the glazing assembly and not just the glazing material itself.

The performance characteristics of glass therefore will vary with its composition. In order to achieve levels of fire resistance from glass, it is necessary to ensure that it is glazed and framed in ways that are specified for that particular products.

Normally, glass can be divided into the following categories:

6.6.1 Non-insulating glass products

These are glass products able to resist the passage of smoke, flames and hot gases, but not able to satisfy the insulation criterion. Regulations may place limits on the location or areas of non-insulating glass that may be used.

There are basically two types of glass which are considered non-insulating:

(a) **Wired glass:** on exposure to fire, the glass breaks due to thermal shock but the wire mesh within the glass maintains the integrity of the specimen by holding the fragmented pieces in place. Wired glass is generally 6mm thick and is manufactured by sandwiching an electronically welded steel mesh between two layers of molten glass in one continuous rolling process.

Wired glass will crack very early in a fire but the steel mesh sandwiched in the centre of the glass holds it together if it has been broken by impact or by thermal shock. Because the integrity and stability are retained in a fire, the spread of smoke and flame is prevented even though the glass may be badly damaged.

(b) **Special composition glass:** on exposure to fire, the glass does not break owing to its low coefficient of thermal expansion, and hence remains unbroken within its frame. The glass may also be thermally strengthened to minimise the effects of stress, thereby achieving a level of impact safety.

6.6.2 Partially Insulating Glass Products

These have fire resistance properties which lie between the insulating and non-insulating glass products. they are usually multi-laminated panes incorporating one intumescent interlayer which becomes opaque on heating. As a result of this intumescent interlayer, they are able to resist the passage of smoke, flames and hot gases and meet the insulation criterion for up to 15 minutes. The temperature on the unexposed surface, after this time, then rises beyond the accepted criterion level, but less quickly than for non-insulating glass.

6.6.3 Insulating Glass Products

These are glass types which are able to resist the passage of smoke, flames and hot gases and meet the insulating criterion for at least 30 minutes. National regulations generally require 30, 45 and 60 minutes compliance.

There are two types of insulating glass available. The first is intumescent laminated glass formed from multi-laminated layers of flat glass and clear intumescent interlayers. The fire resistance depends on the special composition of the interlayers, which react to high temperature by intumescing to produce an opaque shield that resists the transmission of radiant and conducted heat. On exposure to the fire the glass fractures, but remains bonded to the interlayer. The level of fire resistance achieved is directly related to the number of interlayers.

The second is gel interlayed glass formed from a clear, transparent gel located between sheets of toughened glass separated by metal spacer bars and sealed at the edges. The level of fire resistance achieved is related to the thickness of the gel interlayer. On exposure to fire, the gel forms a crust and the evaporating water from the interlayer absorbs the heat energy. This process continues until the gel has burnt through.

6.6.4 Multi-laminate glass

This relatively new fire-resisting glass is manufactured from multi-laminate panes of float glass with clear intumescent interlayers. On exposure to fire the intumescent layers expand to form an opaque shield which forms an effective barrier to smoke and flames and prevents the transfer of radiant and conductive heat.

6.6.5 Heat treated fire glass

This type of glass is wire free and as such can be fractured early on in a fire if subjected to thermal shock. The important aspect here is the 'edge' of the glass. This must be protected from the full force of the fire. It is absolutely vital (as with all windows), however that the glass is installed properly to ensure that it performs correctly and does not fracture.

Glass by itself is not fire resisting. The level of fire resistance achieved is that of the system – glass, beads, glazing materials, frame and frame restraint detail. The whole system is only as strong as its weakest component.

It is imperative however with fire resistant glass that the whole glazing system, including the frame and method of fixing must have fire resistance, not just the glass itself. BS 476, Part 22 gives guidance in this area.

6.7 Insulating materials

6.7.1 Cavities

In order to reduce heat transmission in hollow spaces such as those partitions between an exterior wall and an internal lining, in a floor or in a roof, they are frequently filled with materials which are of a loose fibrous nature and have a low conductivity.

Many substances have been used for this purpose, including such combustible materials as cork, sawdust and peat. Modern research, however, has produced noncombustible substitutes such as rock or glass wool, foamed slag, vermiculite, etc. and these are now replacing the older materials in new buildings, although by no means all insulating materials now being employed are non-combustible (See Figures 6.2 and 6.3).

Polystyrene is well known for its good thermal properties and comes in various forms such as rigid or flexible sheets, in granulated form or as a spray – it has the disadvantage however of having little fire resistance.

To enable them to be laid rapidly in a position, insulating materials are sometimes sandwiched between layers of bituminous paper or felt and are then known as 'quiltings'. Combustible quiltings, e.g. those with a wood or seaweed base, enable a fire to travel easily through concealed wall and ceiling spaces, the plaster or board lining preventing effective extinction until it is removed. This type of insulation is to be found in older buildings and is not a process employed in the construction of today's moden buildings.

6.7.2 Spray-on insulation and intumescent seals

It is essential that the structure of a building is safe under fire conditions and retains its integrity long enough for the brigade to carry out its duties.

Since the advent of sprayed asbestos, whilst this material is generally no longer used, new technology has developed a wide range of spray-on products like vermiculite-cement and sprayed mineral fibre, some of which are designed to withstand high intensity fires which might be experienced, for example, in the petro-chemical industry.

Intumescent strip seals and acrylic mastic have been developed to provide protection and maintain fire resistance in gaps and joints which are flexible enough for structural movement. These various types of internal and external insulation material not only help a building to reduce heat transmission or protect it in the event of fire; they also help to reduce condensation and deaden sound.

6.8 Masonry

Masonry materials have been used for fire resisting constructions ever since the need for fire protection in buildings became apparent. Masonry includes bricks and blocks of clay or concrete covering a wide variety of shapes and densities.

The most common use of masonry is in the construction of walls, of factories, storage buildings, offices and high rise blocks. It can be used in conjunction with concrete, steel or wood structures. Masonry constructions are usually able to withstand exposure to fire without much distress. However, problems arise when other elements, particularly beams and slabs, undergo expansion. A steel beam over a span of 20m can expand as much as 150mm when its temperature is raised by 500°C. This can cause deformation of a masonry wall against which it may be bearing and as a consequence a hole may be punched in the wall or it may be pushed out.

Brickwork is generally a very good fire-resisting material. It is quite possible to achieve periods of resistance of up to four hours, the stability of the material being due to the high temperatures to which it has already been subjected during manufacture. However, there may be problems with large panels (over 4 metres) of brickwork due to differential expansion and movement. In these instances, the restraints being applied to the edges of the panels become critical e.g. brick panels in concrete frame buildings.

Clay bricks are made by the firing of the clay at high temperatures. This imparts to the brick an ability to withstand exposure to fire conditions without suffering much physical damage. They can be considered virtually inert at high temperatures. Failure of brick walls in fire has resulted when the construction could not withstand thermal movement because of its large size and lack of provision for stability and expansion, or when there was movement of other parts of the construction.

The behaviour of solid concrete blocks is similar to that of concrete walls as the material properties are similar, but the presence of mortar joints allows improved capacity to compensate for unequal expansion on the section when one face is exposed to heating. Both hollow and solid blocks are not susceptible to damage by spalling. Aerated concrete blocks, owing to the thermal properties of the product, provide good fire resistance as loadbearing and non-loadbearing systems.

Figure 6.2 *Polystyrene insulating sheeting placed between the inner wall and the brick cladding.*
Photo: Essex Fire Brigade.

Figure 6.3 A specially designed cladding system fitted to an inner hollowbrick wall.
Photo: Langley London Ltd.

6.9 Paint

Paint is used both as a preservative and as a decoration principally for woodwork, steelworks and plaster, and is sometimes applied to brickwork. Paint consists of a pigment (normally a powdered solid) carried in a vehicle (a liquid) which, by chemical action and evaporation, allows the film to harden. Almost all paints (with the exception of fire-retardant paints mentioned later) are flammable, but the film ordinarily is so thin that it has no appreciable effect on a fire, although under certain conditions, it can foster surface spread.

Sometimes, however, when many coats of paint have been applied over a number of years, the film may be sufficiently thick to become flammable and constitute a fire risk. The paint on steelworks, for instance, can ignite if heated sufficiently by a fire, e.g. the far side of a bulkhead in a ship fire. Where appearance is unimportant, tar or bituminous paint is sometimes used and the film may then be sufficiently thick to burn even though applied to steelworks with no other combustible material present.

Fire-retardant paints are occasionally used to protect timber and are of two types. One type is a fairly heavy-based paint which will not inhibit combustion completely, but will do much to reduce flaming, whilst the other, which is termed 'intumescent paint', will, when subjected to heat, bubble up and form a layer of air cells which acts as an insulation between the heat of the fire and the timber underneath. This type of paint is very effective and can be obtained in colour or as a transparent covering.

The development of intumescent coatings proved to be a technological breakthrough. Originally, this type of paint was designed to retard flame spread but has now been developed to react chemically to heat exposure by undergoing a physical change. On reaching temperatures in excess of 200°C (Paint Research Association figures), the paint develops into a thick insulating foam which can protect steel for long periods of time depending on the thickness of the coatings.

Research has shown that multi-layered paints in communal areas and escape routes can represent a significant hazard in terms of flame spread and toxic fumes emission in the event of a fire. Such finishes are often found in circulation areas and fire escape routes within high rise housing block environments which are themselves prone to high levels of vandalism, graffiti and arson.

The paint finishes contained in the circulation area where fatal fires have occurred varied considerably in terms of type and loading. They range from multi-layered paints in excess of 1mm, conventional oil based finishes measuring less than 0.5mm and a finishing layer of anti-graffiti lacquer measuring less that 50 microns.

Poor adhesion of existing finishes is also thought to have had an adverse bearing on the spread of flame and yet, from one notable fatal fire, even a well adhered finishing lacquer appeared capable of promoting unexpected and rapid spread of flame throughout several floors.

From the evidence available, it is suggested that all existing paint films, of unknown origin, and of whatever thickness, age or condition should be perceived as potentially flammable. Unless designed with fire inhibiting properties, all paints in the dried film state contain materials that are combustible. To apply them on top of the other without regard to their combined flammability is to worsen areas already likely to be at risk. Tests have shown that a cigarette lighter held to paint films, even though the finish is firmly adhered to the wall, is capable of igniting them and they will continue to burn after the lighter has been removed.

At the design stage and prior to occupation of residential blocks, there is an obligation to incorporate adequate fire precautionary measures to prevent the unrestricted spread of smoke and fire. There is little doubt that when these buildings are first occupied the fire precautions, including the provision of wall and ceiling surfaces having an acceptable class of flame spread, are of a high order. There has been a commonly held belief that when such buildings are redecorated – which tends to be a regular event as a result of graffiti attempts – the standard of protection afforded the occupants is not reduced in any way.

This situation was first highlighted when the multi fatality fire at Kings Cross Underground Station was investigated. Findings at the time pointed to the effect of the painted ceiling above the escalator aiding the rapid fire propagation. London Underground Scientific Services Division carried out experiments which showed that self-sustaining fire can develop on paint fibre which meets certain criteria. Subsequent investigations into further fires where there was rapid spread of fire and fatalities, showed a significant contributory factor was the type and conditions of the wall linings used: specifically the substrate preparation, the nature and thickness of wall linings, and the condition of decorative coatings applied prior to the finishing coat.

In 1992, the Home Office gave guidance in a 'Dear Chief Officer Letter' to all fire authorities, this guidance was based on work undertaken by the Fire Research Station on behalf of the London Borough of Southwark. In Scotland, similar guidance was issued under cover of Dear Firemaster Letter 1/1992.

This guidance advised that wherever possible prior to redecoration, all existing surface coatings be removed and the non-combustible substrate exposed.

A proprietary coating, for which the manufacturer can provide a certificate showing the material to have obtained a class 1 surface spread of flame rating when tested in accordance with BS476:Part 7, should then be applied in accordance with the manufacturer's recommendations. Ideally the certificate should relate to the material under test on a non-combustible substrate of a similar type to the intended application.

It should be noted that applying a certificated product over existing decorative coating will nullify the claims of the certificate.

It is realised that stripping the paint may be an onerous and costly task and more realistically a Brigade may wish to consider whether a survey of the total fire hazard in a communal area should be carried out. If so, the survey should address the following factors and their bearing on the potential fire hazard:

(a) History of vandalism;
(b) Presence/storage of rubbish within the staircase;
(c) Provision/absence of fire safety arrangements;
(d) Nature of the wall finishes;
(e) The type of substrate.

Damaged or loose coatings may make a greater initial contribution to the fire than those in good condition. Areas of poor adhesion should be removed prior to redecoration. Unsound areas may not always be apparent from visual inspection alone.

6.10 Plastics

Plastic materials fall into the general class of materials known as synthetic polymers – organic substances of high molecular weight, man-made from repeating units of lesser molecular weights called monomers

Reference has already been made to 'plastic boards' but, in recent years, there has been a considerable increase in the use of plastics in building construction.

The term 'plastics' is a generic name for a group of materials based on synthetic or modified natural polymers which at some stage of manufacture can be formed to shape by flow, aided in many cases by heat and pressure.

Plastics has become the most versatile and widely used class of materials known to man. The range of processing and performance characteristics, and of combinations of these characteristics, made available through judicious design has brought these materials into almost every conceivable application. Many of these applications involve a significant possibility of exposure to fire.

Because the essential ingredient of any plastic is an organic substance, combustion can result under sufficiently severe exposure to heat and oxygen. The severity of exposure required to produce combustion and the results of that combustion process, vary as widely as the materials themselves.

Plastics or polymers are available in a variety of physical shapes. The physical form in which the plastic is present has significant influence on its flammability characteristics. In some cases physical structure is more important than chemical structure where fire behaviour is concerned.

Polymers are divided into three classes, which include: elastomers, thermosets and thermoplastics. In addition to the polymer, finished plastics may contain plasticizers, colourants, fillers, stabilisers and lubricants.

An increasing use of reinforced plastics for structural applications has focused increasing attention on reinforcing materials and their performance under fire conditions. There is probably no other domain, regarding the testing of construction materials and assemblies, where the caveat that these items should be tested in environments and under physical conditions matching their actual use is more critical.

The use of plastics in construction materials and assemblies for walls, roofs, and flooring as well as for a wide variety of finishes, furnishings and other appliances which find their way into buildings, has become a major concern to fire protection engineers.

The diversity of the types and formulations in plastics in use is broad, with these materials displaying an extremely wide range of properties and attributes relating to ignitability, flame spread, heat release, smoke production and toxicity. How they are manufactured and conditioned, the type and effectiveness of flame-retardant treatment they undergo, and whether they are used within or on the exterior of a building have important impacts on their performance under fire conditions.

In addition, the fire performance of coatings and membranes used to protect these materials, and thermal barriers used to separate them from interior spaces, are critical to their safe use in building. Finally, the quantities used and their installed configuration can have significant effects on their behaviour in fire.

Plastics can be THERMOSETTING, i.e. they will not soften significantly on heating to a temperature below decomposition temperature, or THERMOPLASTIC, i.e. capable of being softened by the application of heat.

Plastic materials of different degrees of stiffness are described as rigid, semi-rigid and non-rigid plastics. Reinforced plastics consist essentially of polymer combined with fibrous material to enhance its mechanical strength. This term is most commonly used for thermosetting polyester resin with glass fibres (GRP). One of the uses of this material is as external cladding in the form of moulded panels in building systems. It can be formed to a wide variety of shapes, colours and textures; components made from it are light in weight and its mouldability allows the incorporation of detail that would be impossible to achieve with other materials.

Cellular plastics are made up of a mass of cells in which the matrix is a plastics material.

Foamed plastics are cellular plastics made mainly from liquid starting materials, e.g. polyurethane foam.

Expanded plastics are made by stamping or cutting plastic sheet and stretching to form open meshes, in the same way as expanded metal is formed.

The problems of tensile strength and compressive strengths of these materials for their possible use as structural elements have not yet been fully resolved and, except for small complete structures, they are not used for loadbearing members. A substantial amount of plastics material will, however, be encountered within buildings in the form of thermal insulation, service pipes, wall, floor and ceiling covering, furniture, furnishings and fittings. Translucent pvc sheeting is widely used for roof lighting and clear acrylic resins used for shaped lighting panels as in domes. Plastics materials cover such a wide range of substances that their properties and behaviour in fire can be described only in very broad terms. It depends upon the composition and method of manufacture, the free access to air and any support to combustion that may be available. The products of combustion of many plastic materials may be very toxic; again this is dependent on the type of plastic and the combustion of other materials that may be involved in the fire.

The face of construction is changing however with the first "system-built" house in Pittsfield, USA, constructed in the main, of plastic around a timber-framed structure. "System-built" construction is already well accepted in Japan and Scandinavia and, with the rising price of labour, may soon find its way to Europe.

6.10.1 Cellular Plastics

The term cellular plastics covers a wide range of products used in an even wider range of applications. It includes flexible, semi-rigid and rigid foams; materials made in large blocks (e.g. polystyrene) and others made in small discrete shapes. The cell sizes range from below 1mm to 10mm or more.

The main types of polymer used are polyurethane (PU), polystyrene (PS), polyethylene (PE) and polyvinyl chloride (PVC or uPVC). The largest uses for these materials are as packing, in upholstered furniture and for insulation. Although the physical properties of these materials may seem diverse, manufacturing techniques have some common features, and create related hazards. Most cellular plastics can be considered fire hazards.

Many cellular plastics may be ignited with a small ignition source, such as a match. Once established a fire will grow rapidly, often producing a large volume of thick, black toxic/irritant smoke.

All smoke contains toxic and irritant features, but the smoke from some burning cellular plastics may be particularly dangerous. For example, polyurethane releases hydrogen cyanide while isocynates and acrolein is often present on the smoke from burning plastics. There is also concern that some fires involving plastics may produce small quantities of effluent that are not easily dispersed in the atmosphere and may damage the environment if allowed to enter water courses etc.

6.10.2 Polyvinyl Chloride (PVC)

Because of the widespread use of this particular class of plastics, it is considered appropriate to offer additional, specific comment here.

Polyvinyl chloride (PVC, uPVC or vinyl) materials or products tend to have good fire performance properties, in particular, pure PVC will not continue to burn once the source of heat or flame is removed. This is because nearly 57% of the base polymer is chlorine and it is well known that chlorine is one of the few elements that confers good fire properties to a polymer.

Unplasticized (rigid) vinyl materials, such as those used for making pipe, sliding or vertical blinds, present less of a hazard than traditional materials such as wood. When PVC is plasticized to make flexible products such as wire coatings, upholstery or wall coverings, its fire properties become less favourable depending on the amount and kind of plasticizer and other additives used. However, most plasticized PVC products in use still tend to have good fire performance, particularly if additionally treated with fire retardants.

When measured by standard tests, PVC is among the materials with the lowest rate of heat release. The major combustion products given off by PVC are the same as those produced by wood or most other combustible materials, both synthetic and natural. In common with most other products, when burning PVC will give off carbon monoxide (CO); the one product given off by PVC that is not given off by natural materials is hydrogen chloride (HCl) a highly irritant gas.

The data shows that although the mechanisms of action of CO and HCl are totally different, their lethal doses are very similar. Hydrogen chloride has one important feature related to fire hazard – a very pungent odour, detectable at a level of less than 1 ppm, while CO is odourless and narcotic.

6.11 Steel and other metals

Whilst a number of different metals are used to some extent in building, only iron and steel are normally used for those parts which have to carry any load. Cast-iron possesses relatively little strength in tension, but is capable of sustaining a considerable load in compression. Cast-iron was very widely used in the 19th century for beams and columns, and even now, there are many buildings in use which are supported by cast-iron columns and beams. Iron and steel used in the construction of a building are not combustible and present no risk of fire-spread from direct burning.

Unprotected metal surfaces may, none the less, constitute a serious risk in a fire because all metals heat up and expand when exposed to fire and are also a potential cause of fire-spread by conduction. Unprotected metal which is used to carry a load also presents the even more serious danger of rapid collapse when excessively heated. Structural steel, for example, loses two-thirds of its strength at 593°C and, in proportion to the amount and direction of the load to which it is subjected, begins to sag and twist. This is by no means an abnormal temperature in even a moderate fire – the danger of the failure of unprotected loadbearing metal work cannot be over-emphasised.

Steel is one of the most important metals used in the construction industry. Without it, construction would be limited to massive all-masonry buildings.

There are two distinct ways in which steel is used in building construction – either as a structural material or as reinforcement in concrete. In the first case the steel sections may be of substantial size, rolled into various shapes and connected to steel or concrete elements. As reinforcement, the material is in the shape of bars, wires or tendons.

Steel is very strong, its modulus of elasticity (m), a term which measures its ability to distort and restore is about 29m psi – far more than any other material. Steel's compressive strength is equal to its tensile strength. Its shear strength is about 75% of its tensile strength. This great strength enables steel members of relatively small mass to carry heavy loads, particularly when used in trusses.

Fire resistance, however, is a function of mass therefore strong but lightweight members have little inherent fire resistance. The effect of fire on steel can be crucial to the stability of almost any building during a fire.

Unprotected steelwork will lose approximately half of its strength in temperatures of 500-550°C and is therefore very vulnerable in a fire. As a result it is essential that steel structural assemblies are protected either by insulating materials or by the dissipation of the heat on the steel.

There are a variety of insulating materials for steelworks. Other structural materials (e.g. brick and concrete) can be used, but this is a very expensive solution. The more common materials are insulating boards, sprayed coatings or intumescent paints. Insulating boards can be used to encase steel beams and columns. There are also available insulating sheet materials which can be used to protect whole walls. The technology of their use is well documented, but care must be taken in the detailing of all junctions to ensure that no areas of unprotected steelworks are exposed. The disadvantage for the designer is the added bulk which this encasement of steel means, and the care which must be taken to ensure proper construction. Sprayed coatings are normally of mineral fibre or vermiculite cement. Intumescent materials react to heat by expanding and forming an insulating layer.

They can be applied to steelworks as sprays or paints and have the advantage of retaining the profile of the structural element. The disadvantage of intumescents lie in the more limited length of fire resistance they can provide (normally one hour, as opposed to a maximum of two hours for sprayed coatings and four hours for boarding), and in the susceptibility of the material to abrasion damage either during construction or during the lifetime of the building. Intumescents must also be applied over suitably cleaned and primed steel, and the thickness of the layer must be checked before a top sealer coat is applied.

Dissipation of the heat away from the steel is a more exotic option, but it is possible to use water to cool hollow steel sections or flame shields to reduce heat gain. It is even possible to design the structure such that it is outside the building envelope and therefore protected from the risk of internal fires. However, these techniques are very expensive and require maintenance throughout the life of the building.

A 10 metres steel joist, for example, will expand 60mm for a 500°C rise of temperature, and where it is built into a loadbearing wall, such expansion may cause collapse. In a framed building, the failure of a single beam or column is unlikely to cause more than a local collapse. It is clear then that all structural steel must be protected either by solid or hollow protection.

Because it is prone to rapid corrosion, stainless steel is invariably coated, normally with a protective metal coating in the first place, and, outside this, frequently with some form of mineral and resin mixture. This is liable to produce the anomaly that, whereas steel itself is non-combustible, a protected steel sheet of the type described may well have a surface spread of flame of Class 1, 2 or even 3.

Increasing use is being made of aluminium and its alloys for structural and cladding members and this has created new fire problems. The advantages of using aluminium alloy in buildings are:

- A reduction in the weight of the structure;
- Resistance to corrosion;
- Ease of handling and working; and
- The high strength to weight ratio.

The disadvantages are:

- The very rapid loss of strength in fire (stability is affected at 100°C to 225°C);
- The high expansion rate (approximately twice that of steel);
- Its very low melting point (pure aluminium melts at 660°C).

6.11.1 Lead, Copper and Zinc

Lead was principally employed for internal plumbing, flashings and roof coverings. It melts at 327°C and precautions should therefore be taken against injury from molten metal when working beneath a lead roof at a fire in older buildings. Copper and zinc are also used for roof coverings but their melting points are much higher and the metal usually oxidises away under the influence of the fire so that there is rarely much danger from falling molten material. Bronze has a melting point about 1000°C, but is normally only used for decorative grilles, handrails, etc. and occasionally for window frames.

6.12 Stone

Natural stone used for building construction falls within the following general classes:

(a) Igneous rocks – these result from a molten material which subsequently solidified, one of the most common in this group is granite;
(b) Sedimentary rocks – these result from a breakdown of igneous rock into small particles by the changing climatic conditions. The particles fall into low-lying areas to form layers which over very many years become hardened by pressure. Among this group is sandstone and Portland stone;
(c) Metamorphic rocks – these result from igneous and sedimentary rocks which have undergone a natural change by pressure or heat, and in some cases, both. Amongst this group is marble.

The types of stone principally employed in building are granite, sandstone and limestone. Igneous rocks, such as granite, contain free quartz, which has the peculiar property of expanding very rapidly at 575°C and completely shattering the rock. Considerable spalling at the surface may occur in a fire and thin sections of stone may disintegrate entirely. Limestones are composed principally of calcium carbonate, which decomposes at about 800°C into free lime and carbon dioxide. The change is gradual with little alteration in volume, and as heat is absorbed in the process, the interior of a block of limestone may be protected by the outer skin. Water used in firefighting will slake away the quicklime so formed and will cause the outer skin to fall away.

A recent fire in a Victorian building involved the collapse of a limestone staircase in unexpected circumstances. Scientific and structural analysis of the collapse revealed that the apparent cause did not comply with the known facts. The water applied to the staircase did not cause any spalling of the limestone and the temperature at which the limestone is recognised to decompose was never reached.

Sandstone generally comes between granite and limestone in fire behaviour and may shrink and crack in a fire.

Stone is, in general, a good heat insulator, but is inferior to brick when subjected to continuous heat, because of its tendency to spall or split into pieces, especially when water is suddenly applied. Stonework should always be carefully watched for signs of cracking when it is necessary to work beneath or near it.

Stone and granite is now normally used for facings and decoration and its failure in the event of a fire is unlikely to affect the stability of the structure.

6.13 Timber

Wood products constitute one of the most important groups of materials used in construction. No other construction material is so prominent in international trade.

Wood is a combustible material which consists essentially of cellulose, hemicellulose and lignin. It is a non-homogeneous and variable material, and even for a given species the properties (particularly structural properties) are dependent upon the grain direction and the size and nature of discontinuities such as knots, shakes etc.

Solid sections of common structural timbers can be used for beams of widths up to 100mm or for column up to 150mm square. For larger sizes the use of laminated sections allow timber to be used to best advantage, as these have more predictable structural properties and can be fabricated to a given length or span.

It is therefore common to find solid timber sections employed in the construction of beams for floors, columns and framed walls for use in domestic buildings, offices, schools etc. With few exceptions, for these constructions timber sections are used in conjunction with various lining materials. Timber floors and timber stud walls are therefore examples of composite constructions in which the combined behaviour of timber and the associated materials determines structural behaviour in fire.

Although wood is associated in many countries with fire risk, attitudes are inconsistent. There are in fact wide differences in national building codes and regulations and in other factors governing the use of wood products in building.

Wood burns, but because it burns at a regular, measurable rate it is possible to deliberately oversize timbers, so that they can be used as structural elements. Such oversizing is often described as 'sacrificial timber'.

The surface degradation of the wood is normally in the form of charring, and the flaming will occur only with temperatures at the surface in excess of 350°C and the presence of a pilot ignition source. As the outer surfaces of a timber member char, they tend to stay in place and the inner core of the wood remains relatively unaffected and can retain its stability and integrity. The rate of charring may vary from 0.5mm min-1 (oak, teak) to 0.83 mm min-1 (western red cedar), but a value of 0.67 mm min-1 is a widely accepted estimate for structural species.

This approximation applies both to solid members and laminates, though laminates may actually perform better as they will not be so prone to knots or other deformations of the timber. The use of flame-retardant treatments will not normally slow down the charring rate. It is, of course, possible to protect timber by the use of insulation materials, but as the choice of timber has probably been made because the designer wishes it to be exposed, this is an unattractive option in new buildings. However, it may be necessary when improving the fire safety of existing timber structures to consider the cladding of timber elements with insulating materials.

The great advantage of timber to the designer is that failure is predictable and will occur slowly, the great disadvantage is the dramatic increase in cost of timber elements which have had to be deliberately oversized.

Loadbearing timber walls are usually found in low rise buildings intended for domestic or office accommodation Consequently the fire resistance requirements are for 30 or 60 minutes depending upon the use of the wall. Such walls are an essential component of timber-framed housing where the walls consist essentially of a timber stud framework of storey height, 2.8m or so, in which the uprights are held in place by timber plates at the top and bottom with one or more stiffeners or noggins.

Although there exist examples of timber-framed houses dating back to Tudor times it would be fair to say that the vast majority of houses constructed in timber in this country have been built since 1964–65. Because this form of construction enables a wide range of claddings to be used (e.g. brick outer skins, tile hanging or cement rendering on metal laths), in addition to timber –

which itself can be fixed to otherwise traditional brick houses – it is not always obvious whether or not a house is timber-framed.

The period mentioned began with the introduction, for the first time, of national building regulations. These, together with accumulated knowledge from other countries having a long and continuing history of timber dwellings, have resulted in standards of construction being adopted in Britain which make the modern timber-framed house once completed, with its fire protective linings, as safe as the more traditional forms.

For centuries timber has commonly been used structurally for floors, roofs, beams and columns, internal partitioning and staircases. More recently it has been used in the construction of a wide range of buildings and structures from homes to leisure centres and bridges.

As one would expect, timber is combustible and there are various flame retardant treatments that can assist in making wood more difficult to ignite. There are two types of flame retardant treatments for timber:

(a) Surface coatings – painted onto the timber surface with little or no penetration into the wood. Some doubt exists as to the permanency over a period of time.
(b) Impregnation – using a combination of pressure and vacuum this process 'drives' the liquid into the timber. This is a type of permanent treatment but it may affect the cosmetic appearance of the timber in moist atmospheres.

However, the performance of timber in real fires is frequently far superior to unprotected, non-combustible materials such as steel and aluminium for the following reasons:

(i) Timber does not expand significantly under the influence of heat (in fact, it may shrink slightly) and buildings reliant upon timber for structural purposes are not likely to suffer sudden collapse brought about by 'unrestrained' expansion.
(ii) British Standards 476: Part 5 comes to the conclusion that timber in sizes normally used for construction purposes is defined as '...not easily ignitable'.
(iii) Timber has the inherent ability to protect itself; the build-up of charcoal on the surface of burning timber limits the availability of oxygen thereby insulating the remainder of the section.
(iv) It has been established that the burning or charring rate is predictable and varies only slightly with species of timber and not on the severity of the fire. 'Sacrificial' timber built into the construction may be consumed by a fire before the structural core is attacked.

Laminated timber has become popular in Britain and has been used for decades in Scandinavian countries, as well as France and Germany. This type of timber keeps its structural integrity and its cosmetic appearance is one of its attributes. Spectacular spans of over 150 metres can be achieved thereby reducing the requirement for connectors which can represent a weak link in any structure.

Chapter 7 – Examples of buildings

7.1 Introduction

Having described materials and building components in previous Sections, in this Section an attempt is made to explain how they come together, with fire safety in mind, in the design of buildings.

Building Regulations stipulate standards of construction and are aimed at the safety of people and of enhancing their chances of escaping from a building on fire without injury. They also seek to prevent or limit fire-spread inside a building and from one building to another.

The increasing complexity and size of buildings today where, in some cases, literally thousands of people work or resort to, poses many problems. An example of this is a town centre development. This may include department stores, shops, offices, hotels, car parks, restaurants, theatres, a leisure centre, indoor and outdoor markets and multi-level walkways. To make this type of environment safe yet workable architects, building control officers and fire prevention officers etc., have to plan carefully.

Because of their good record so far, there is little information as to how one of these complexes does react to a fire. However, fires have occurred and have highlighted some problems e.g., reverse flow of smoke, failure of smoke detectors and/or sprinklers to operate due to unforeseen smoke or heat barriers. This has caused a rethink in design and, in some cases, alterations to legislation and practical guidance, which, in turn will affect future designs.

Large complexes and buildings are often now designed as "intelligent" constructions. This means, usually, that there is a central control room, which monitors all the systems in the building including those exclusively used for fire safety.

It must be borne in mind that quite a few buildings are designed for "general" use and it is the occupier who decides on its final internal layout. Alternatively, buildings are designed for a particular use but if the first occupier moves out, the use changes but the "particular" design does not. This can cause problems for the fire prevention officer and the firefighter. In the Building Regulations the purposes for which buildings are used (or intended to be used in the case of proposed buildings) are divided into "purpose" or "occupancy" groups. These groupings reflect the potential hazard that is expected to exist in relation to the safety of people likely to resort to the premises, and to the fireload i.e., the amount of combustible material present. At the top of the groupings are residential and institutional buildings, which have a sleeping risk. In a lot of cases this situation is worsened by the fact that some occupants find it difficult to move without assistance.

The scale continues through to buildings used for storage where the fire damage potential may be high but the life risk is relatively low. All groups take safety of people into consideration. Amongst many things, the regulations limit the size of the buildings in relation to the period of fire resistance of its elements of structure. Having determined the potential hazard of the use of the building, the Regulations seek to ensure that it has sufficient fire resistance to prevent:

(a) Major collapse and
(b) Large scale fire.

In this Section, examples of buildings will be given according to these groupings but, as there are so many variations in each group, only a representative selection is made. The Regulations apply to:

- New buildings;
- Structural alterations or extensions to existing buildings (irrespective of when the building was erected);
- Certain works or fittings (including replacements) e.g., drains, incinerators;
- When a material change of use is deemed to have been proposed.

Firefighters still have to consider buildings erected before the introduction of the Regulations, which are, often, comparatively sub-standard to those erected now. Examples are given, as far as possible, of both "old" and "new" buildings used for similar purposes

7.2 Residential and institutional buildings

In England and Wales, the standard of fitness for human habitation is defined in Section 604 of the Housing Act 1985 as amended by Paragraph 83 of Schedule 9 of the Local Government and Housing Act 1989. This fitness standard is not a comfort standard but is related to the health and safety of the occupants of the housing accommodation concerned. The fundamental concept of the new standard is different from previous standards in that there is no longer to be a cumulative or combined approach to the various standards. The requirements listed in Section 604 of the 1985 Act must all be met for premises to be fit for human habitation.

The Fitness Standard addresses fire safety as part of this remit when, in the opinion of the local housing authority, a dwelling (single occupancy and multiple occupancy housing) fails to meet one or more of the following requirements which place the occupants at risk from fire-related harms:

(a) It is structurally safe;
(b) It is free from serious disrepair;
(c) It has adequate provision for lighting, heating and ventilation; and
(d) There are satisfactory facilities in the dwelling house for the preparation and cooking of food, including a sink with a satisfactory supply of hot and cold water.

Guidance for surveyors on the application of the new standard makes reference to 'fire' in several action areas including freedom from serious disrepair:

> 'The disrepair of fixtures can also be seriously prejudicial to safety, either directly or by constituting a fire hazard.'

As well as causing deaths directly through electrocution, bad wiring results in numerous house fires each year. Old and neglected wiring, particularly the rubber-covered cable used up to the 1950s, is more likely to be faulty and cause fires. The disrepair of gas-fired boilers, space and water heaters may also cause a fire hazard or may lead to the emission of toxic gas and vapours, the latter resulting in many accidental deaths in the home each year.

7.3 Terraced and semi-detached houses

Older types of terraced housing have the outer walls and the separating walls of brick or stone. The internal walls are more commonly of wooden studding faced each side with lath and plaster. The internal walls often carry part of the weight of the floors which span between them and either the outer or separating walls. The wooden floor joists have boarding on top and lath and plaster or plasterboard ceilings underneath. The roof is usually constructed of timber and is tiled or slated. Virtually the whole interior of this type of house is constructed of timber and fire-spread can be rapid.

These houses are often sub-divided into flats and tenements, the resulting overcrowding increasing the risk to life. Means of escape can often be inadequate and access to the rear of the building difficult, if not impossible.

Modern terraced housing, sometimes of 3 or 4 floors, is now called "town-housing". The ground floor is often a garage with the living rooms on the first floor and the bedrooms above. Construction varies but floors of 22mm sheet plywood on timber joists are not unusual plus bituminous felt on timber flat roof. In most cases the imperforate separating walls are taken through the roof by at least 375mm to prevent fire-spread between occupancies.

A problem arises in old terraced housing where, sometimes, the separating walls are only carried up

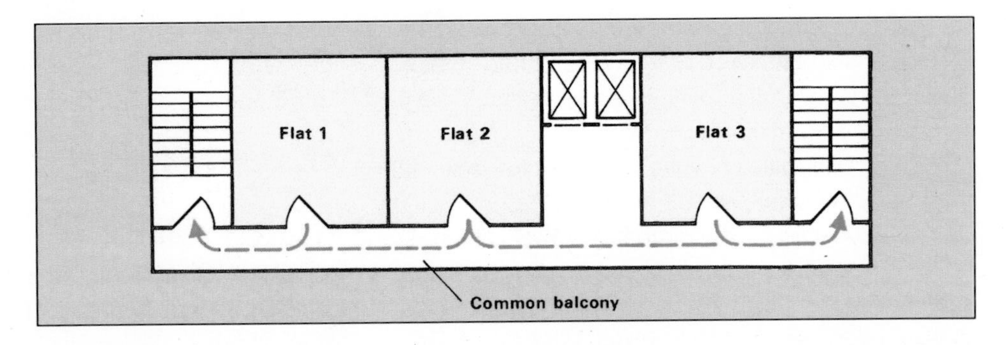

Figure 7.1 Layout showing a number of maisonettes with a common balcony escape route to staircases at the end.

to the top ceiling level leaving a "common" roof void running the whole length of the terrace. If fire penetrates the top ceiling anywhere in the terrace it can spread through the roof space and down into other occupancies.

Semi-detached houses of the older type can have the same problems because of similar construction. Roof construction in modern types has become very light in cross-sectional area and any involvement can lead to rapid spread of fire and collapse.

7.4 Flats and maisonettes

7.4.1 Flats

For all practical purposes the term "flat" means a dwelling forming part of a large block with common access which has all its habitable rooms and kitchen on one level or, in the case of "split-level flats", not more than half a storey apart. Purpose built flats are, generally, of more than two storeys and are referred to as "multi-storey", "high-point" or "tower blocks".

7.4.2 Maisonettes

A maisonette differs from a flat in that, although it forms part of a larger block with common access, its habitable rooms and kitchen are divided between two, or more, levels which are more than half a storey apart.

The design of maisonettes varies greatly. There are the simple type approached from a common open balcony or deck with living rooms at entrance level and bedrooms on the floor above.

Firefighters, however, will find more complex interlocking forms through two or more floors with living rooms and bedrooms on any floor and possibly two entrances at different levels.

7.4.3 Means of escape

BS 5588 Part 1 specifically deals with the protection of the occupants of flats and maisonettes, the means of escape, the construction of the building and the equipment necessary for preventing the rapid spread of fire.

The concept of the use of manipulative apparatus for means of escape or the external rescue using Fire Service ladders is no longer acceptable. The protection must be designed and built into the building with adequate fire resistance between dwellings, limits on travel distances, alternative means of escape etc.

Figure 7.1 gives a simplified example of alternative means of escape from a flat. Any route from the staircase to an external door at ground, or podium, level is regarded as an extension of the staircase and must be protected accordingly.

7.5 Houses in multiple occupation

In England and Wales, under the Housing Act 1985 the local authority is obliged to require means of escape in case of fire in certain types of houses which are occupied by persons not forming a single household (Houses in Multiple Occupation). A house in multiple occupation is defined in section 345 of the Housing Act 1985

However, compliance with the guidance in Approved Document B – Fire Safety, will enable a newly constructed or converted House in Multiple Occupation to achieve an acceptable standard of fire safety.

In Scotland under Sections 162 and 166 of the Housing (Scotland) Act 1987, a local authority has power to require such works it considers necessary to provide means of escape from fire in houses in multiple occupation. The local authority is required to consult the fire authority before using this power. Separate, discretionary powers to introduce houses in multiple occupation licensing schemes were given to local authorities by the Civic Government (Scotland) Act 1982 (Licensing of Houses in Multiple Occupation) Order 1991.

There are a number of other Statutes enforced by the local authority or the fire authority that may be applied to premises of specific uses once they are occupied.

Guidance is also available in:

- Department of the Environment. Houses in multiple occupation. Guidance to local housing authorities on standards of fitness under section 352 of the Housing Act 1985. Circular 12/92.

The circular gives guidance to local housing authorities in England on the standard, which they might consider adopting when exercising their powers under this Act. The provisions for means of escape from fire and other fire precautions replace "Guide to means of escape and related fire safety measures in existing Houses in Multiple Occupation 1988".

- Welsh Office. Local Government and Housing Act 1989. Houses in multiple occupation standards of fitness Circular 25/92.

The circular gives guidance to local authorities in Wales on the standards which they should consider adopting when exercising their powers under section 352 of the Housing Act 1985. The provisions for means of escape from fire and other fire precautions replace "Guide to means of escape and related fire safety measures in certain existing Houses in Multiple Occupation, 1988".

In Scotland, The Scottish Home and Health Department Fire Service Circular 2/1988 "Guide to means of escape and related fire safety measures in existing homes in multiple occupation in Scotland" also provides advice. This guide deals with certain types of houses in multiple occupation and hostel type accommodation (including common lodging houses).

7.6 Timber-framed houses

Many old timber-framed houses remain and are usually constructed of stout timbers with an infix of brick noggin or plaster. Internally the floors are often of heavy timbers as is the roof, which can be tiled, slated or thatched. The numerous concealed spaces brought about by centuries of alterations make any fire in this type of building particularly difficult and dangerous.

Timber-framed houses have, in the last 20 years, experienced resurgence but the construction is very different.

The main framing is of light timbers with very light roof trusses. A great deal of prefabrication is used and the trusses are, sometimes, of only 40mm finished thickness and they are not nailed or screwed but held together by different types of metal connector plates (see Figure 7.2).

The house frame walls are completed by prefabricated panels of plasterboard sheathed with plywood or fibreboards. These, in turn, are covered with a membrane, which acts as a breathing weather protection. This membrane can be made of chemically waterproofed Kraft paper or bitumen-treated paper. As both of these can be easily ignited the tendency is to use thermoplastic sheeting which will shrink away from a heat source instead of igniting. The whole frame is then clad in brickwork.

Ground floors usually consist of a concrete slab with timber or plastic flooring laid directly onto it. There is a move back to the raised type of ventilated timber ground floor with services running under. Upper floors can be of 12–13mm planks on rafters or, in some cases; plywood sheeting on timber rafters, both underdrawn by plasterboard ceilings.

Firefighters should be aware that some ceilings might conceal a heating system. This could be a low temperature radiant system where the heating elements are built into the ceiling often by stapling the current-carrying metal foil to the underside of the joists or battens.

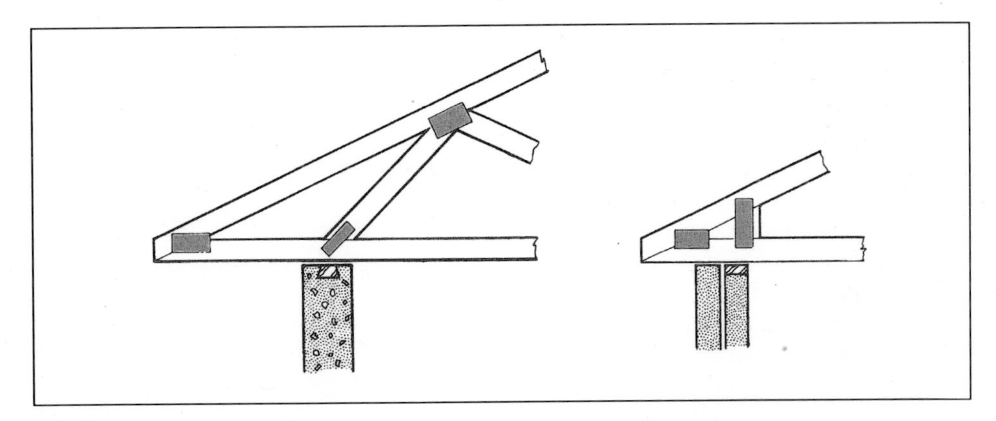

Figure 7.2
An illustration of how metal connectors are used to join roof members.

In the one system the metal foil is embedded in a strong plastic sheet, the total thickness being about 0.2mm. The elements provide about 200W/m^2 from the normal voltage of 220/240. Plasterboard would then be erected in the normal way leaving no obvious indication of the presence of the heating system.

Modern installations should be provided with an Earth Leakage Circuit Breaker (ELCB) or a Residual Current Circuit Breaker (RCCB) which are to set to trip out if the system is damaged.

Firefighters should question the occupant if they have any suspicion that this sort of system is fitted and take all the normal precautions as when dealing with live electrical circuits.

Insulation, both in the roof and walls, is currently mineral fibre (rock or glass) but expanded polystyrene and expanded polythene have been permitted.

From a fire-spread point of view, providing the appropriate fire-stopping has been correctly installed and the construction is good, fires can be contained. Unfortunately, however, this type of construction does engender numerous cavities and, if the fire breaks through into this, spread can then be rapid and largely unseen.

The large amount of lightweight timber surface area in the roof also means a rapid spread if the fire penetrates into the roof space. Another factor to be borne in mind is that the metal fasteners/connectors have been known to expand and fall out of the timber trussing in a fire situation leaving the clay tile roof unsupported and causing rapid collapse. Firefighters should be aware of this trend in construction and be wary of climbing onto any of these roofs which have suffered internal fire damage.

7.7 Hotels

Hotels come within the "other residential" user group of buildings for the purpose of the Building Regulations and it is these regulations which limit the floor area and cubic capacity of each storey or compartment.

In general, the design of hotels is such that the lower floors contain the amenities e.g., ballroom/ conference area, bars, restaurant and kitchens. Upper floors are given over to bedrooms and suites.

These arrangements depend often, of course, on the peculiarities of that particular site. The advent of atria, large expanses of glass or polycarbonate, roof gardens, energy conservation ideas etc., has lead to completely new designs (see Figures 7.3 and 7.4).

In some buildings the traditional position has been reversed with the amenities on the upper floors and the suites below. This can add to the problems of means of escape whereby the largest numbers of people will be the farthest from safety and has led to a great deal of new thinking on the interpretation of the Building Regulations.

The Fire Precautions Act, 1971 was enacted to enable fire authorities to require reasonable measures to be taken in hotels, and similar premises, to protect people using these premises. The "Guide to fire precautions in existing hotels and boarding houses" has been issued aimed at those premises

Figure 7.3 Interior of an ultra-modern hotel. The central lift shaft can be seen left. Note the fire detector left of the sculpture and unopenable windows to the atrium. Photo: *Heathrow Sterling Hotel.*

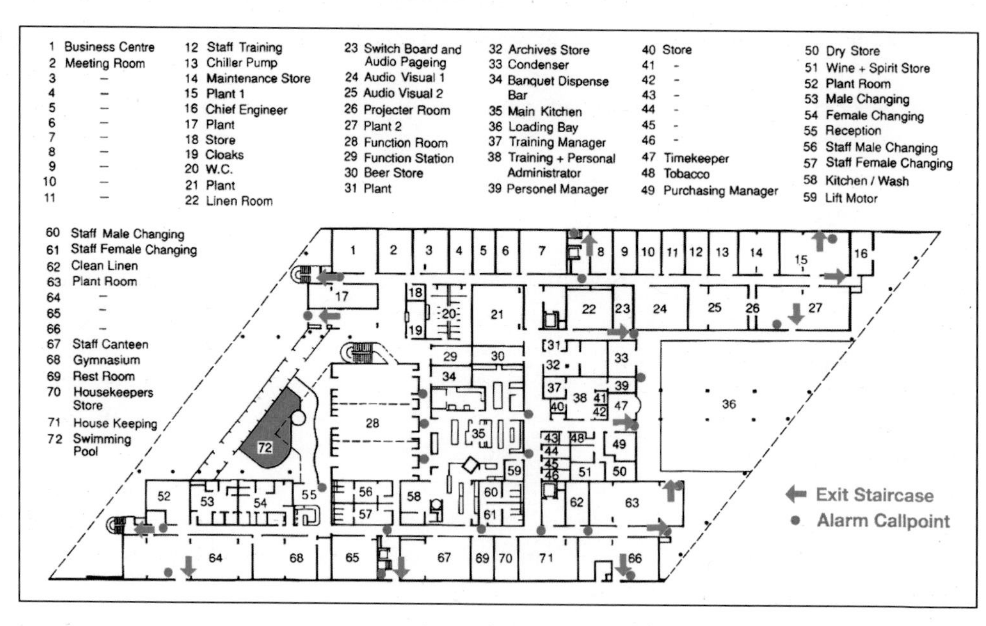

Figure 7.4 An example of a floor plan in a large modern hotel.

requiring a fire certificate. This was necessary because the Fire Safety and Safety at Places of Sport Act, 1987 amended the 1971 Act giving additional powers to fire authorities and additional requirements.

To assist owners and managers of these types of premises, the Home Departments, together with the Fire Protection Association, has issued a guide "Fire safety management in hotels and boarding houses".

Firefighters must bear in mind that the majority of people in hotels are there for the first time, are almost totally unaware of the general layout of the floors, exits etc., do not absorb the printed fire instructions in their rooms and are generally dis-orientated, at least for a time. Under fire conditions this can be, and too often is, fatal.

The design of hotels, the fire prevention measures incorporated and the general requirements of the

1971 Act all attempt to redress this situation but the more the local firefighter knows about the premises the greater the chance of success in any firefighting or rescues.

7.8 Institutional buildings

7.8.1 General

This designation includes buildings used as hospitals, community homes, boarding schools and similar establishments where persons in need of care sleep on the premises. No specific type of building covers these establishments which can be converted country mansions; large private houses quite often used as private nursing homes; single and two storey modular built units used as old persons' homes right through to multi storey, steel-framed hospitals.

7.8.2 Fire precautions

The greater life-risk in these premises is recognised in the Building Regulations and the requirements for new and converted buildings are very stringent.

7.8.3 Fire Safety in Healthcare Premises

A fire in a hospital, healthcare or social services premise would pose a major threat to the lives of everybody within it but particularly to the patients or residents. Such premises, therefore, require a fire safety strategy based primarily on avoidance of fire. In the event of fire there must be means for rapid detection, containment and control, supported by reliable and rehearsed procedures for removing patients or residents to places of safety.

7.8.4 Firecode

Firecode is a suite of documents that provides guidance on a wide range of issues that affect the provision of healthcare by NHS Trusts (HSS Trusts in Northern Ireland). Firecode is applicable in Northern Ireland and takes into account the administrative and legislative obligations under the Health and Personal Social Services (Northern Ireland) Order 1991 which corresponds with the National Health Service and Community Care Act 1990.

A similar suite of guidance documents has been published for use in Scotland by the NHS in Scotland Estates Environment Forum under the title "NHS in Scotland Firecode".

The guidance is contained in Health Technical Memoranda (HTM) which are general, wide ranging documents, supplemented as appropriate by occasional issues of specific information relating to specific identified problems; and in Fire Practice Notes (FPN) which are intended to address specific issues. The issue of NHS Trust Estate Published Letters (EPL) may deal with other points.

The guidance may also be used as good practice guidance for fire risks in non-NHS hospitals and nursing homes registered under Part II of the Registered Homes Act 1984, the Registered Homes (Northern Ireland) Order 1992, the Residential Care Homes Regulations (Northern Ireland) 1993 or the Nursing Homes Regulations (Northern Ireland) 1993.

The Health Technical Memoranda are usually prepared by a working group chaired by NHS Estates, in consultation with the Home Office, Department of the Environment, Department of the Environment (Northern Ireland), CACFOA, Fire Research Station and the National Association of Hospital Fire Officers. Other publications, such as the Fire Prevention Notes are usually prepared within NHS Estates and may be subject to consultation before publication.

The Home Office accepts that the Firecode suite of documents contains the Department of Health's policy and technical guidance to health authorities on fire precautions in hospitals and other NHS premises. They contain agreed national guidance on fire precautions in hospital premises, including those owned and managed by NHS Trusts.

Firecode documents are distributed to all fire authorities in England and Wales, by the Home Office under the cover of a Dear Chief Officer Letter with the recommendation that each section should be used with all other appropriate sections, when advice is sought from fire authorities. The Department of the Environment for Northern Ireland and the Scottish Office Home Department have endorsed this recommendation.

The advice may be that sought under Section 1(1)(f)

of the Fire Services Act 1947 or Article 4 of the Fire Services (Northern Ireland) Order 1984, both of which place a statutory duty on the Fire Authority to advise on fire precautions when requested to do so. This duty is unaffected by the provisions of the National Health Service and Community Care Act 1990 and the Health and Personal Services (Northern Ireland) Order 1991.

The Secretary of State for Health retained powers of Direction under the National Health Service and Community Care Act 1990, which requires NHS healthcare premises, including Trust premises, to comply with the provisions of Firecode. In particular, it should be recognised that HTM81 – Fire precautions in new hospitals satisfies all the requirements of Part B of Schedule 1 of the 1991 Building Regulations, and the Building Regulations (Northern Ireland) 1994.

NHS in Scotland Firecode is recognised by The Scottish Office as giving guidance on good practice for compliance with statutory and mandatory requirements particular to healthcare premises. The guidance does not, however, override any statutory provisions and it should be noted that some of the guidance might require the relaxation or waiver of Technical Standards to the Building Standards (Scotland) Regulations 1990.

When using Firecode, it is important to recognise that it is not possible to make comprehensive recommendations covering all eventualities and it should always be borne in mind that the purpose of hospitals is to provide medical treatment and/or nursing care. The complex nature of hospitals will sometimes require a more flexible approach to ensure that the correct balance is achieved between fire safety and the requirements for treatment and nursing care.

This should be done on the basis of professional judgement and an understanding of the nature of the problems. However, care should be taken to avoid compromising the safety of patients, visitors and staff.

One fundamental parameter must be recognised and that is in the design of hospitals, no reliance is placed on external rescue by the fire brigade or manipulative types of escape appliance such as chutes or portable ladders. The principle is that in an emergency the occupants of any part of a hospital should be able to move, or be moved, horizontally, to a place of relative safety with the assistance of staff only. Vertical movement is seen as only being necessary as a last resort.

All the Firecode guidance has been prepared on the understanding that it will be used by competent persons who have sufficient technical training and actual experience or technical and other qualities, both to understand fully the dangers involved and to undertake properly the measures referred to in the documents. Fire officers of local authority fire brigades fall within that definition provided they have had sufficient technical training.

The particular problems of demands for increasingly high levels of security in patient areas (both preventing patients from leaving and preventing unauthorised entry by persons, particularly in child and maternity units) has raised a number of issues. The possible conflict between patient security and means of escape in case of fire can be complex.

The use of premises as a hospital is therefore not a use designated for the purposes of Section 1 of the Fire Precautions Act 1971. A Fire Certificate issued by the fire authority however, may be required for those parts of hospitals, which are used as shops, offices or factories and fall within the numerical provisions of that Act.

7.8.5 The Department of Health Nucleus Hospital

The Nucleus hospital is an assembly of standard cruciform templates each of which may be arranged horizontally and vertically to form a low-rise hospital with landscaped courtyards providing good natural light and ventilation.

The required standards of fire safety for this unique system of building, engineering and operational policies were agreed with the Home Office (type approval) and published as 'Nucleus fire precautions recommendations'. This document is used for fire risk assessments in Nucleus hospitals. However, any work affecting fire safety in these hospitals should follow the guidance. At no time should HTM86 – Fire risk assessment in hospitals be used to relax the Nucleus standards. Reference should also be made to Estate Published Letter EPL (97) 5. It should be noted that the 'Nucleus' hospital planning guidance did not apply to Scotland.

The Nucleus scheme was classified as conforming to the Nucleus concept by submission to the Department of Health and the Home Office. Fire authorities should be aware that there are fundamental differences in the provision of the Nucleus guidance and that contained in either HTM81 – Fire Precautions in new hospitals or HTM85 – Fire precautions in existing hospitals. If the integrity of the design provisions is to be maintained, alterations or additions to Nucleus hospitals should be in accordance with the Nucleus fire precautions recommendations and not HTM81 or HTM85.

One of the most common suggestion puts to fire authorities is that the courtyard space should be infilled with other structures. Such provision would compromise the integrity of the Nucleus concept and should be resisted. Other difficulties have been experienced at the interface of a Nucleus scheme with an HTM81 or HTM85 hospital, particularly with the provision of safety lighting and fire alarm systems.

NHS Estates announced in an Estate Published Letter (EPL 96/3) that no more hospitals would be designated as Nucleus. However, there were a considerable number of such hospitals built in the United Kingdom and it is possible that some NHS Trusts will, at some time in the future, seek to extend them, either retaining the Nucleus concept or introducing schemes that will not. In the latter case, the full provision of HTM85 should be applied.

7.8.6 Health Care Buildings with an Atrium

An atrium within a building can provide an exciting multi-storey space allowing natural light into a deep plan building. The architectural potential of atria has long been recognised in hotels, offices and shopping complexes. Hospitals are now adopting a more market oriented approach with the development of Trusts and the role of purchasing introduced by the National Health Service and Community Care Act 1990. Within this increasingly competitive environment, the use of atria to enliven hospitals by creating stimulating spaces to attract health purchasers and to improve income generation is expected to increase.

The use of such building techniques presents the fire safety engineer with additional problems.

Fire Practice Note FPN8 – provides guidance on the additional fire precautions required when atria are incorporated in the design of an NHS hospital. These are required because atria provide a route by which fire and smoke may spread more rapidly than in an equivalent non-atrium hospital.

The guidance in FPN8 supplements that contained in:

(a) Guide to fire precautions in existing places of work that require a fire certificate;
(b) Code of practice for fire precautions in factories, offices, shops and railway premises not required to have a fire certificate; and
(c) Approved Document B, or Parts D and E of the Technical Standards in Scotland.

It should be read in conjunction with these three documents.

7.8.7 Commercial enterprises on hospital premises

Fire Practice Note FPN5 – Commercial enterprises on hospital premises provides general technical guidance relating to the additional fire safety precautions which may become necessary when commercial enterprises are about to be, or have been, established on hospital premises.

It is intended that the recommendations of FPN5 should be applied to commercial enterprise areas formed by the conversion, adaptation, extension, modernisation or refurbishment of existing locations within, or closely adjacent to, hospital buildings. In the case of new hospitals, the recommendations should be considered at the initial planning stage.

Such recommendations cannot take account of all of the circumstances which may be found in any particular hospital, but are intended to highlight the health service aspects which need to be considered. The guidance is meant to ensure that when commercial enterprises or complexes are planned or designed, they will not subvert the fire safety precautions already agreed for adjacent health care areas or hospital buildings.

The note requires consideration, *inter alia*, of the need for additional fire detection, alarm and extinguishing systems and smoke control systems beyond those normally provided for hospital premises

7.8.8 Patient Hotels

Patient Hotels are intended to provide good quality accommodation for low-dependency patients ("guests") who are mobile and able to look after themselves, and consequently do not require the full facilities of a hospital's acute ward. Typical guests would be:

- Those requiring diagnostic procedures;
- Day patients or out patients who could not be sent home for reasons of distance or domestic circumstances; and
- Those whose condition does not require nursing care for all or part of their stay.

Throughout Fire Practice Note 7, people staying in patient hotels are referred to as "guests"; this reflects the aim of patient hotels, which is to provide a level and quality of service comparable to a hotel.

The following facilities would not normally be provided in a patient hotel:

(a) Medical gases and other bedhead services, although emergency alarms may be provided to call nursing staff or medical assistance from elsewhere in the hospital;
(b) Nursing or medical care.

Patient hotels, which provide any of the facilities, listed above, fall outside the scope of this FPN and should be designed to follow the guidance contained in Firecode HTM81 – Fire precautions in new hospitals.

Many of the "guests" who fall within this category may be elderly people or people with a disability. They may be slow moving and require additional time to escape in case of fire. To address these concerns, NHS Estates has prepared Fire Practice Note FPN7. This Note suggests that patient hotels should be provided with higher standards of detection and alarm and means of escape than would normally be provided within a hotel.

Fire Practice Note 7, therefore, provides guidance on the standard of fire safety applicable to:

(i) Purpose-built patient hotels, either as parts of NHS hospital premises, or on a separate site;
(ii) Conversion of non-ward hospital accommodation into patient hotels.

Where it is proposed to convert existing patient care areas of hospitals into a patient hotel, the effect of the conversion on the overall standard of fire safety within the hospital needs to be considered. The major use of the building will still be a hospital and consequently, to ensure that the overall fire safety standards are not compromised, the conversion should be considered as a "major alteration", as defined in HTM81 and should be designed to comply with the guidance in that document.

Where patient hotels are established within a "template" of a Nucleus hospital, they should be designed to comply with the guidance in 'Nucleus Fire Precautions Recommendations'.

Patient hotels are not necessarily subject to the Fire Precautions (Hotels and Boarding Houses) Order 1972. Whether or not they require a Fire Certificate – in whole or in part – is a matter initially for local determination on the facts of the case and ultimately for the Courts to decide. Home Office has issued guidance (Dear Chief Officer Letter 10/1995) setting out its view that may assist in the determination. In Scotland reference should be made to Dear Firemaster Letter 4/1996, item A.

Any arrangement that is confined to the provision of sleeping accommodation for NHS patients would not appear to fall within the use designated by the 1972 Order, whether the premises are run by NHS management or by commercial contractors. It also seems that an arrangement which includes the provision of sleeping accommodation for other restricted classes of "guest" (such as relatives of patients and visiting healthcare professionals) is not likely to make the use a designated one, even if such guests have to pay for their accommodation. However, any patient hotels which are commercial enterprises providing sleeping accommodation for members of the public at large would require certification if the number of people exceeded the numerical threshold for application of the Order.

The Fire Practice Note supplements the guidance in the 'Guide to Fire Precautions in Premises used as Hotels and Boarding Houses, which require a Fire Certificate' published for the Home Office and the Scottish Office. The FPN also supplements the Department of the Environment's Approved Document to Part B of the Building Regulations 1991 (in Northern Ireland it is the Building Regulations (Northern Ireland) 1994). For the purposes of the Building Regulations it has been agreed with the Department of the Environment that patient hotels should be classified as Purpose Group 'Residential – 2(b)'.

The FPN should, therefore, be read in conjunction with these two documents. The Home Office further recommends that it should be used by fire safety officers in conjunction with the existing guidance when responding to requests for advice.

7.9 Educational Buildings

7.9.1 Constructional Standards for Schools

Most building projects in England require approval under the Building Regulations, made under the Building Act 1984. But building projects at most schools are exempt from the Building Regulations, provided that the Secretary of State has approved particulars of the project. For building work at grant-maintained and at grant-maintained special schools, approval must be sought from the Funding Agency for Schools (FAS).

In Scotland, educational buildings are required to comply fully with the requirements of the Building Standards (Scotland) Regulations 1990. Supplementary guidance is provided in Educational Building Note No. 18 'A guide to fire safety in schools' published by the Scottish Office Education Department.

Section 4 of the Building Act 1984 (as amended by paragraphs 86 to Schedule 19 to the Education Act 1993 and paragraph 59 of Schedule 37 to the Education Act 1996) says:

"Nothing in this part of this Act with respect to building regulations, and nothing in any building regulations, applies in relation to:

(a) A building required for the purposes of a school or other education establishment expected to be erected according to:

    (i) Plans that have been approved by the Secretary of State,
    (ii) Particulars submitted and approved under section 39 or 44 of the Education Act 1996 or under regulations made under section 218(7) of the Education Reform Act 1988,
    (iii) Particulars approved or adopted under section 214, 262 or 341 of the Education Act 1996, or
    (iv) Particulars given in a direction under section 428 of that Act."

The Department for Education and Employment (DfEE) interprets the Building Act 1984 (as amended) to mean that schools falling within the scope of their guidance may not seek approval of building projects under the Building Regulations in place of approval from the Secretary of State.

Building works in the following:

- County Schools;
- LEA Special Schools;
- Voluntary Aided Schools;
- Special Agreement Schools;
- Grant Maintained Schools;
- Grant-Maintained Special Schools and
- Non-maintained Special Schools,

therefore, do not have to meet Building Regulations. Instead, they must gain the Secretary of State's approval that they meet the standards set out in non-statutory documents known collectively as Constructional Standards for school buildings.

The DfEE produces similar guidance for the designers of school buildings that fall outside the scope of their guidance.

Constructional Standards contain standards, *inter alia*, for fire safety and access for disabled people. The latest document was issued as a DfEE Circular to replace the current Constructional Standards that were published in 1985.

The question of the continued exemption of school building work from the requirements of the building regulations is being considered. Over a number of years Government policy has brought the requirements for schools buildings gradually closer into line with those for most other buildings. As a result the current standards are already quite close to Building Regulations requirements, and those in the latest documents bring them even closer.

The need for the continued exemption of school building work to remain outside the control of Building Regulations may no longer be appropriate.

The logical end point of this process would be the ending of exemption for school buildings, and bringing school building work under the umbrella of the Building Regulations.

The removal of this exemption (Section 4 of the Building Act 1984 (as amended)) would probably bring additional responsibility on Fire Authorities under the existing consultation arrangements with building control.

It is recognised by DfEE that while it is desirable to use Approved Documents as the basis for the approval of school building projects wherever appropriate, the Approved Documents do not adequately reflect the needs of schools in all respects. As well as referring to certain Approved Documents, the constructional standards continue to specify certain requirements, specific to the school context, with which the Secretary of State expects proposed building projects to comply. However, it is hoped that the closer alignment with Approved Documents should make the task of planning school-building projects more straightforward.

The Constructional Standards for Schools Building Projects, contains those parts regarded as overriding variations to the Approved Document Part B. The DOE revision of Approved Document B now contains almost all of the departures from agreed standards. In addition, in the revision of Approved Document B, the section (3.32), which refers to the Department of Education and Science Building Number 7 – Fire and the design of education buildings, has been deleted.

Specific differences introduced into Approved Document B include:

(a) The variation to 3.5(a), increasing the number of occupants up to 60 (plus supervisors) from the maximum of 50 on the basis that this population represents a certain disciplined occupancy under the direct control of responsible staff;
(b) The proposed travel distances for single direction of travel (18m) and for alternative directions of travel (45m) are the same as shop and commercial premises in 4(b);
(c) The recommended distance in section 3.21 and Table 3, between doors on corridors more than doubles the generally accepted spacing (30m as opposed to 12m).

On the basis of (a) above;

(i) The restriction, Section 4.5(b) of school children to no higher than the first storey in buildings with a single escape stair subject to an overall limit of 120 pupils and supervisory staff provided that the escape route is protected, in addition there are to be no high hazard areas;
(ii) The reduction in escape widths (Table 5) is acceptable on the basis of (a) above. There is also a restriction in respect of spiral stairs;
(iii) Schools or parts of schools used during the normal school day and provided with natural light need not be provided with escape lighting (section 5.33), see also Design Note 17 – Guidelines for Environmental Design in Schools;
(iv) The introduction of the concept of refuge into school buildings;
(v) There is also an amendment to B2, which introduces a more onerous requirement for wall linings in escape routes, acknowledging the tendency of most schools to line corridors with examples of the children's work.

7.9.2 Design Note 17 – Guidelines for Environmental Design in Schools

Section B – Lighting

This section makes no mention of the required design for safety or emergency lighting that may be required if areas require the provision of such lighting as per the Construction Standards for Schools Building Projects. Section 5.33 of Approved document B only states that escape lighting should not be required.

There will also be instances where safety lighting will be required in places of public assembly that most schools have and used for stage plays etc., when members of the public are admitted and a licence may be required under appropriate legislation.

Section C – Heating and Thermal Performance

This section makes no mention of any fire safety provisions in respect of the boiler installation or storage of bulk fuel supplies.

7.9.3 Access for People with disabilities to School Buildings

It must be recognised that there are dangers in talking about "disabled people" or "the disabled". It is therefore more responsible to speak of "a person with a disability".

Section 8 of the document covers risk assessment, means of escape and the management implications of means of escape strategies. This section introduces BS5588: Part 8 – Code of practice for means of escape for people with a disability, and the concept of refuges. With more and more children with disabilities being integrated into schools as part of the education process, such provision is necessary if these children are not to be restrained in their use of the facilities. The management requirements are clearly identified in this section and include consideration of visitors with disabilities to school premises.

7.9.4 Guidance also available is:

Department of Education and Science. Fire and the design of educational buildings. Building Bulletin 7, 1988.

The bulletin is issued in support of the Department's Constructional Standards for building work at educational establishments. The recommendations apply to new construction but may also be appropriate in the context of adaptation or remodelling work to existing buildings. Deals with means of escape, precautions against fire, structural fire precautions, fire warning systems and firefighting by the occupants, and fire prevention and damage limitation.

Department of Education and Science. Safety in science laboratories. DES Safety Series No 2.

Section 2 of the bulletin 'Laboratory design and furniture' includes recommendations for structural fire precautions and means of escape; Section 4 'Fire' covers firefighting equipment, fire prevention, accidental explosions and other fire risks.

7.10 Prisons and other Penal establishments

Because of the prevailing level of supervision and control, the risk of accidental outbreak of fire in a penal establishment is low. Nevertheless, the consequences of an incident can be extremely serious. Guidance given here can only be of a general nature since the circumstances at prison incidents will vary and can change very suddenly, even during the course of the incident, due to the nature of the inmates.

The all-pervading nature of security means that special consideration has to be given to both fire safety and firefighting measures. To this end, the Governor of a prison will have an appointed officer responsible for risk assessment, fire planning, fire prevention and drills.

Buildings and their surround are designed and constructed so as to deny internal freedom of movement to the occupants and their access to the outside world. Potential conflict between security requirements and fire related measures could only be resolved by close liaison between prison authorities and fire brigades.

7.10.1 The Fire Services Act 1947

The provisions of the 1947 Act do not apply to prison establishments, as they are Crown Property.

Fire Officers in charge at prison incidents do not therefore have their normal statutory powers under the 1947 Act. This makes very little difference in practice as prison authorities are generally happy to co-operate and the senior fire brigade officers will be in full command of firefighting, rescue and evacuation procedures within the emergency area.

If, on making the initial attendance security/safety reasons dictate that fire crews are not permitted into the establishment, officers should consider what preparation can be carried out to speed deployment when clearance is given. It is not the role of the fire service to assist in quelling riots in prisons, nor should fire service equipment be used for this purpose either by fire service personnel or prison staff and equipment should not be loaned to prison officers for firefighting.

The following paragraphs highlight some of the area firefighters need to be aware of, they cover features of construction, security, the people and their activities, firefighting systems and firefighting procedures

7.10.2 Features of construction

The buildings of some penal establishments may date from the Napoleonic and Victorian eras; others may be converted old 'mansion' type properties and, if listed buildings, may have restrictions on the extent and type of fire protection measures that can be installed.

Prisons (including those contracted out, i.e., privatised) built in recent years will contain many modern fire related features, but it must be remembered that as Crown Property they are all exempt from the requirements of the Building Regulations and the Fire Precautions Act 1971. They will however, comply wherever this is practical.

The general means of escape principle that occupants are able to move away from the scene of the fire and proceed to a place of safety by their own unaided efforts obviously cannot be complied with in situations where movement is necessarily inhibited by locked doors and gates.

As a compensating factor there is strict control over sources of ignition and combustible materials supported by disciplined officers on constant watch who, in turn, may be assisted by fire detection and suppression systems linked to an on-site, 24 hour manned, control room.

7.10.3 Security

All penal establishments will exercise control over the movement of persons and vehicles, both in and out of the site, including fire brigade resources. Various forms of detection may be employed on both visitors and inward vehicles, e.g., x-ray machines, full body detectors and body searches. Levels of internal security vary considerably according to the category of the establishment, which reflects the type of inmates that it holds.

7.10.4 Work processes

The work processes likely to be found within a penal establishment may rival those to be found on many modern light industrial estates, with the added complication of being confined within a smaller area. Sizeable offices for prison administration; catering facilities, including kitchens, bakeries and dining areas; and laundries may be found.

Factory type production facilities may include one or more of the following:

(a) Plastic raw materials and plastic goods production;
(b) Printing;
(c) Timber industry and wood working;
(d) Light engineering;
(e) Weaving, knitting and tailoring.

To service the needs of this range of activity, a prison will have extensive storage facilities ranging from local stores containing only a day's supply of materials to large main storage areas covering a wide range of commodities.

7.10.5 Fire safety systems

Hazards from the build-up of smoke and the products of combustion are particularly acute in the closed environment of a prison and are heightened by constraint on the means of escape in the event of fire on the inmates. Smoke control methods may include natural or mechanical ventilation, and containment by smoke doors or pressurisation. Firefighters should be conscious that the benefits of pressurisation might be lost if too many doors are left open in the course of firefighting operations.

The essential aim of fire safety systems in prisons is therefore to provide for:

(a) A reasonable smoke free environment outside the cells enabling prison officers to reach them and carry out the unlocking procedures;
(b) Prevention of horizontal and vertical smoke spread between cells; and
(c) Relatively smoke free routes whereby the occupants can proceed away from the area of the fire to a place of safety.

7.10.6 Category of Prisoner

People detained in Her Majesty's prisons are categorised 'A', 'B' or 'C' with category 'A' being the worst offender. Prison establishments are also categorised according to security standards as 'B', 'C' or 'D'. Dispersal prisons normally contain category 'B' prisoners. Category 'C' prisons are for inmates serving medium- to long-term sentences and category 'D' prisons are for short-term sentences. Prisons may also contain those prisoners on remand, i.e., those awaiting a Court appearance, for whom bail has not been granted, and those awaiting sentencing following a Court appearance. Some prisons may accommodate inmates of only one sex; others are for young offenders (up to the age of 21 years).

7.10.7 Floating Accommodation Units

Moored in ports around the world, ships are being turned into floating accommodation units to house volatile groups such as prisoners.

Floating Accommodation Units are far from new. During the Napoleonic wars, near-derelict hulks of former Royal Navy and Merchant vessels were based at various ports, predominantly in the south-west of England, to house the ever increasing numbers of prisoners taken.

In mainland Europe and Scandinavia, such facilities have been considered and in one German city – Hamburg, a considerable number of these units provide temporary accommodation. One such unit is currently in use at Portland in Dorset.

Assessing on-board fire and security risks

Two key aspects must be considered:

(a) What method of construction has been used? and
(b) Are the materials used in that construction suitable for the levels of fire resistance required by statute in the marine world for the accommodation areas of ships?

Preferably, the materials and methods of construction used should be to the standard specified by the International Maritime Organisation (IMO) in its 'Construction Rules for Passenger Vessels'.

Although no IMO legislation currently specifically mentions floating hotels or accommodation units as such, other bodies would most definitely be interested in the fire and safety standards on board.

It is therefore true to say that the basics of fire safety regarding construction, materials and equipment are in place.

Where the accommodation is to be used as a prison unit, security is obviously of the paramount importance. Both the prison authorities and the staff on site would appreciate this, in exactly the same way as it would be appreciated in shore-based establishments. The potential for arson is probably no greater, provided that adequate control of the residents is provided by staff.

7.10.8 Guidance is available as follows:

Home Office 1994. Police buildings design guide.

Volume 1 of the guide is primarily for use by the police building authority to assist it in formulating its proposals.

Volume 2 is primarily for use by the design team. It includes design principles and standards, and model accommodation levels. Reference is also made to relevant operational, security and safety constraints, and to recommended details of construction and detail design.

Home Office 1989. HM Prison Service. Fire precautions manual.

The manual gives guidance on the fire safety management of prisons.

Home Office 1989. HM Prison Service. Design briefing system PF24 Fire requirements.

The document describes the principles of design for fire protection purposes, and the standards of building projects in penal establishments.

Home Office 1990. HM Prison Service. Fire standards in prison establishments: principles of design and standards of construction.

The document applies to all categories of prison establishments up to four storeys, including new complexes and major refurbishment works in existing establishments.

Scottish Prison Service. Fire Safety: Principles of Design and Standards of Construction.

This document applies to all new prison buildings in Scotland and major extensions, alterations and changes of use to existing buildings.

Scottish Prison Service 1991. Fire Precautions Manual.

This manual gives guidance on the fire safety management of prisons.

The Scottish Office Home Department 1988. The Planning of Police Buildings (Scotland).
The memorandum sets out recommended standards for modern police buildings. Part I of the memorandum contains the design guidance and Part II the related appendices.

7.11 Commercial and industrial stores

7.11.1 Shops and departmental stores

The term "shop" covers, virtually, an unlimited range of premises from the small shop on the corner to the largest hypermarket or departmental store.

The problems created by all but the smallest shops led to the 'Offices, Shops and Railway premises Act, 1963' (OSRA) which gave fire authorities some powers to require a degree of fire protection for the public. New buildings are governed by the Building Regulations which, among other things, aim to restrict the extent of the open floor areas and cubic capacity of the building and/or compartment within the building.

The present standard is directed towards the safety of life by:

(a) Planning escape routes;
(b) Planning to prevent the spread of fire;
(c) Constructing and finishing with non-hazardous materials and embodying adequate fire resistance into the structures;
(d) Segregating the high-risk areas e.g., selling areas from the non-selling areas i.e., stores, loading bays, receiving and despatch departments etc.

For selling, the contents of shops are often displayed in such a way as to enhance fire-spread, impede the movement of firefighters and create the maximum heat and smoke in the shortest possible time. This often delays close tackling of the fire which gives it time to affect the structure.

One aspect of which firefighters should be aware is that many small shops are conversions from former private houses. Placing a steel joist across what is now the shop window area to support the upper walls was often carried out. Frequently it has been found that these are not suitably protected against fire with the possibility of the collapse of the complete front of the building.

Where towns do not have new shopping complexes specially built they sometimes develop old covered markets and glaze over the surrounding narrow streets. Fire protection measures are, obviously, looked at with care but the reaction of these older parts of the town to fire can never be accurately assessed and firefighters, again, should be vigilant when tackling fires in this type of area. These mixed hazards are highlighted in BS 5588.

Figure 7.5 Modern office construction with atrium. Fire detectors are visible at second floor level.
Photo: The Fitzroy Robinson Partnership.

7.12 Offices

This type of occupancy can be found almost anywhere and of any construction often as part of another sort of occupancy. Design guidance for new office buildings or alterations to existing buildings is contained in BS5588 Part 3 and Part 11. These make specific recommendations on fire protection, number and position of exits, prevention of fire-spread and general fire precautions.

New office buildings include, among other things, atria, large areas of glass or polycarbonate, curtain walling, cladding of various types, wall climbing lifts and whole floors given over to services (see Figures 7.5, 7.6, 7.7 and 7.8).

To accommodate the mass of electronic technology many buildings will have "access" floors i.e., raised floors giving access to cables, ducting etc.

Not least important are the numbers of people who work or resort to these buildings. A complete evacuation of a 25/30-storey office block could take an hour so they are designed with compartmentation to a high degree of resistance to fire and smoke spread and protection to the means of escape. Figure 7.9 illustrates the sort of plan, which could accompany a fire certificate for one floor of an office building. The degree of protection is self-evident.

Many office blocks are now designed as "intelligent" buildings e.g., all services controlled by computers with, usually, a control room somewhere in the complex fully manned 24 hours a day. Firefighters will know of this location and be assisted by the numerous visual and audible displays.

A new factor, which is arising, particularly in office blocks, is the provision of access and means of escape for people with disabilities. BS 5588 Part 8, 1988 and the Building Regulations both discuss and recommend appropriate standards. These are

Figure 7.6 Interior of offices under construction (see Figure 7.7).
Photo: Richard Turpin.

Figure 7.7 Exterior of Figure 7.6 showing almost total glazing with services at each end.
Photo: Richard Turpin.

Figure 7.8.a Typical atrium of an office block. All the vertical glass is of the "Pyrostop" fire-resistant type. Photo: Pilkington Glass.

Figure 7.8.b Pyrostop fire-resisting glass showing how the intumescent layers react to heat.
Photo: Pilkington Glass.

additional factors, which the management and the Fire Service must take into consideration.

7.13 Airport complexes

This type of complex is hard to define. The larger airports are so diverse in the facilities they provide and deal with such vast numbers of people that they are commercial, industrial and places to which the public resort together on a grand scale.

They present a particular problems to Fire Prevention Officers with the confluence of people, aircraft, flammable fuel, shops, malls etc., Their design varies but great care is taken to make them as safe as possible within the Building Regulations. Figure 7.10 (bottom) is a cross-sectional drawing of part of Stansted Airport and gives some idea of the public areas. Figure 7.10 (top) is a projection of one of the satellites and shows the movement areas and services.

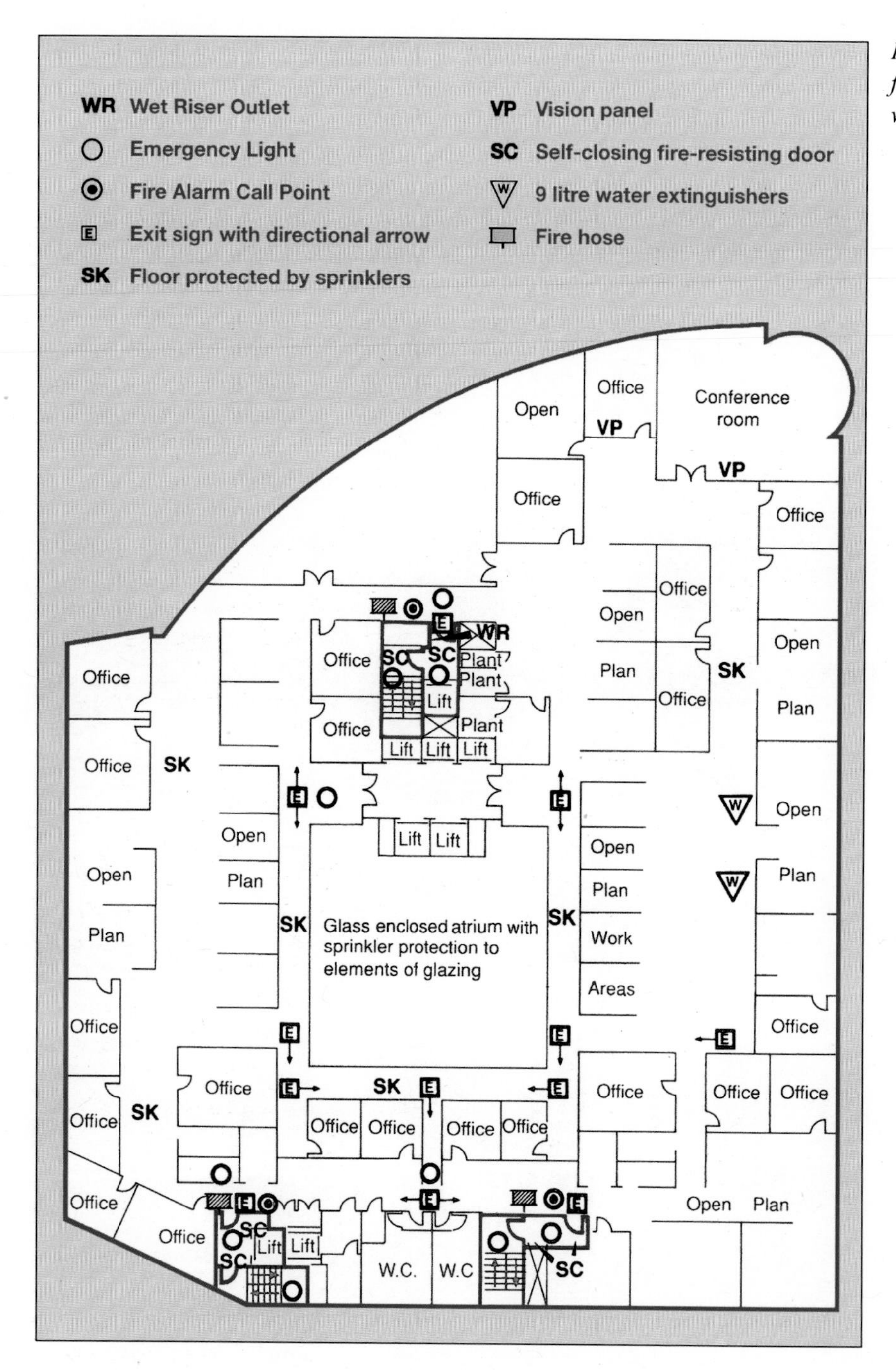

Figure 7.9 Example of the type of floor plan in an office block which would accompany a fire certificate.

7.14 Industrial buildings

During the 19thC a large number of manufactories were built and many are still standing although almost all are not used for their original purpose Figure 7.11 illustrates a typical construction of that period.

Today many new factories appear to be constructed on the single storey principle with large internal dimensions with perhaps a mezzanine floor for various service departments e.g., offices, stores, drawing sections. Some will have these facilities in a 2/3-storey block at one end with a large single storey factory building attached.

Many factories and storage facilities are being built using a light frame of steel or precast concrete covered in steel, or other alloy, sheets. These sheets are the outer skin enclosing a "sandwich" of polyiso-

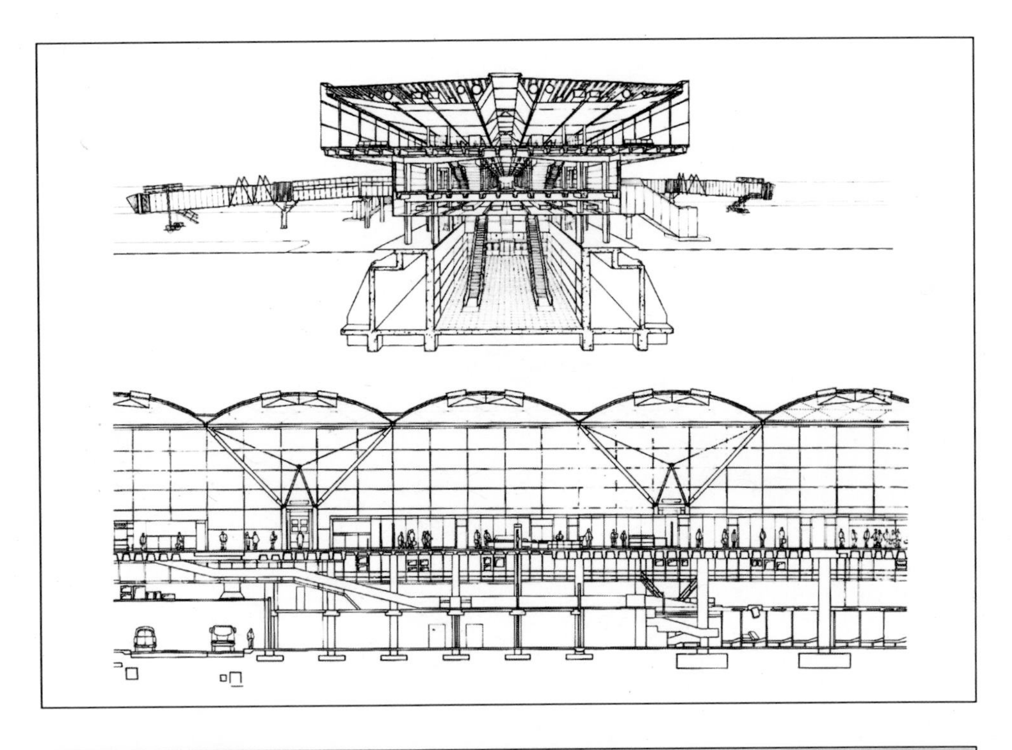

Figure 7.10 (top) An airport satellite area in cross-section. (bottom) Cross-section of part of Stansted airport.

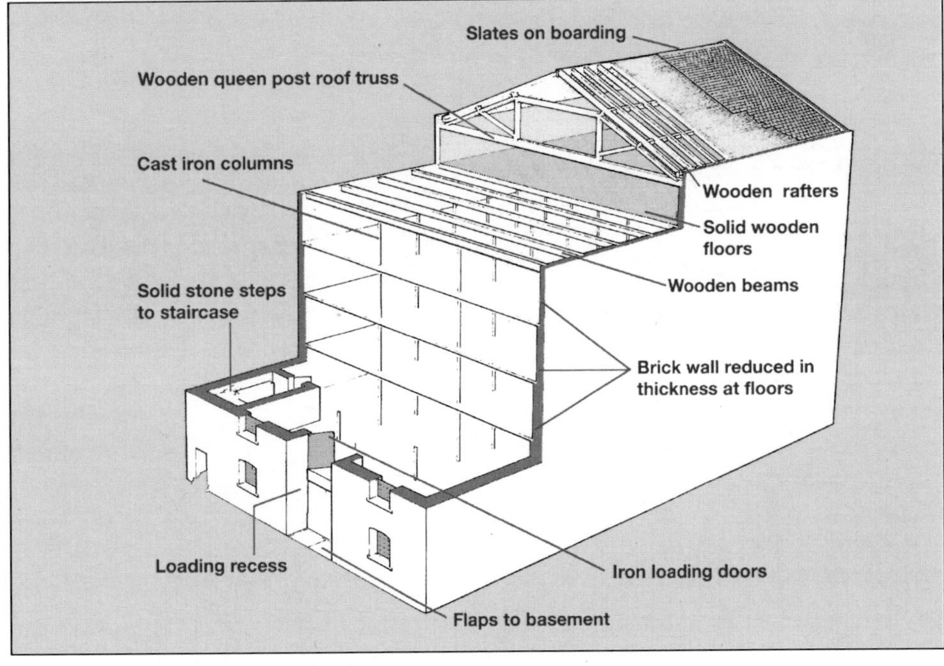

Figure 7.11 Typical form of solid construction with load-bearing walls used during the 19th century for largebuildings such as mills, warehouses, etc.

cyanurate or polyurethane foamed plastic covered internally with either another metal sheet or commercial board (Figure 7.12). These "sandwich" sheets are often specially designed to be attached, in blocks, to the frame and to make a continuous wall and roof (see Figure 7.13), the ends of the building being enclosed by similar sheeting.

Most industrial layouts work on the principle of access for raw materials, processing through the plant with a final destination for finished commodities either into store or onto the dispatch bay for delivery.

Although many of the older multi-storey factories still use the top-to-ground system, the single storey floor system seems to be prevalent.

The dangers of the top-to-ground system are found in the multitude of openings between storeys to

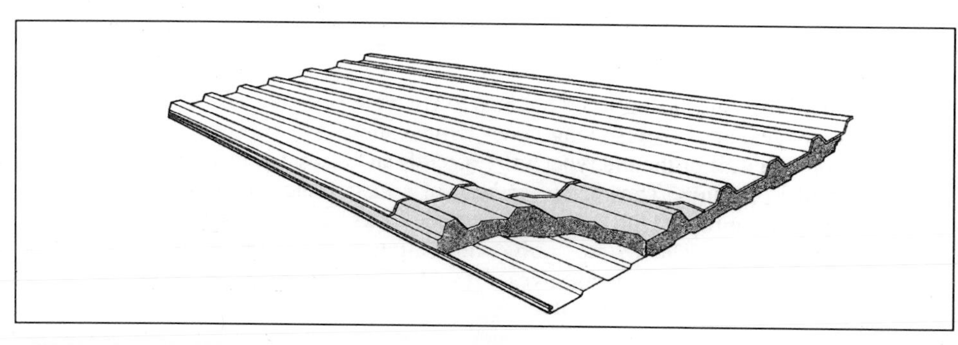

Figure 7.12 A typical "sandwich" sheet used to make up the "skin" of a building.

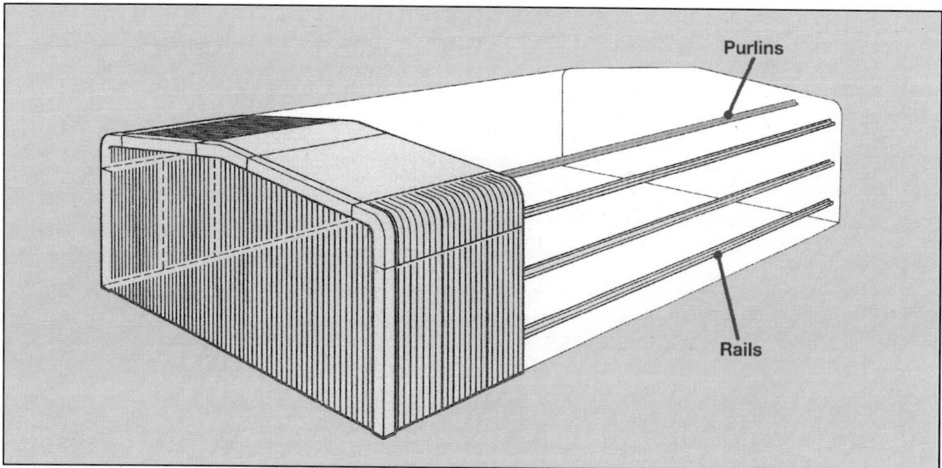

Figure 7.13 Sketch illustrating typical framing and metal cladding.

accommodate various power transmissions, chutes, ducts, conveyors, systems etc. The dangers of the single storey system lies in the, sometimes, vast open areas which are found under one roof an example being in vehicle manufacture. Obviously, high-risk areas such as spray booths and flammable goods stores must be strictly segregated by fire-resisting structures. A simplified layout of such a factory is shown in Figure 7.14.

7.15 Cold stores

These buildings are, usually, particularly difficult for access as their design is geared to preserving low temperatures by substantial insulation (itself, if of the old type, a fire hazard) and restricted internal and external openings. They are often built of a monolithic reinforced concrete frame infilled with brick or concrete blocks, the whole being "double-skinned". These are sub-divided internally into large compartments, which are, themselves, "double-skinned". Additional hazards to firefighters are either the release of refrigerants during a fire or the spread of fire within the double skin. These, and other salient points are discussed in the Manual Part 6c, Section 12 "Refrigeration plant risks".

7.16 Spirit Storage

A particular example of a storage building requiring special consideration is that containing whisky in bulk. Guidance has been issued on the design of new single storey buildings used for spirit storage in the Fire Prevention Guide No. 2/1973. Attention is drawn to:

(a) Limitation of compartment size;
(b) Light roof construction to vent any explosion;
(c) Installation of a sprinkler system;
(d) Restriction of racking height;
(e) Specific amounts of natural ventilation;
(f) Use of approved intrinsically safe electrical equipment.

7.17 Automated and high-bay warehouses

7.17.1 General

Large stores, super- and hypermarkets and other retail outlets need a sophisticated retail distribution organisation. The requirement to store huge quantities of goods which have a rapid turnover has led to the development of very large **A**utomated

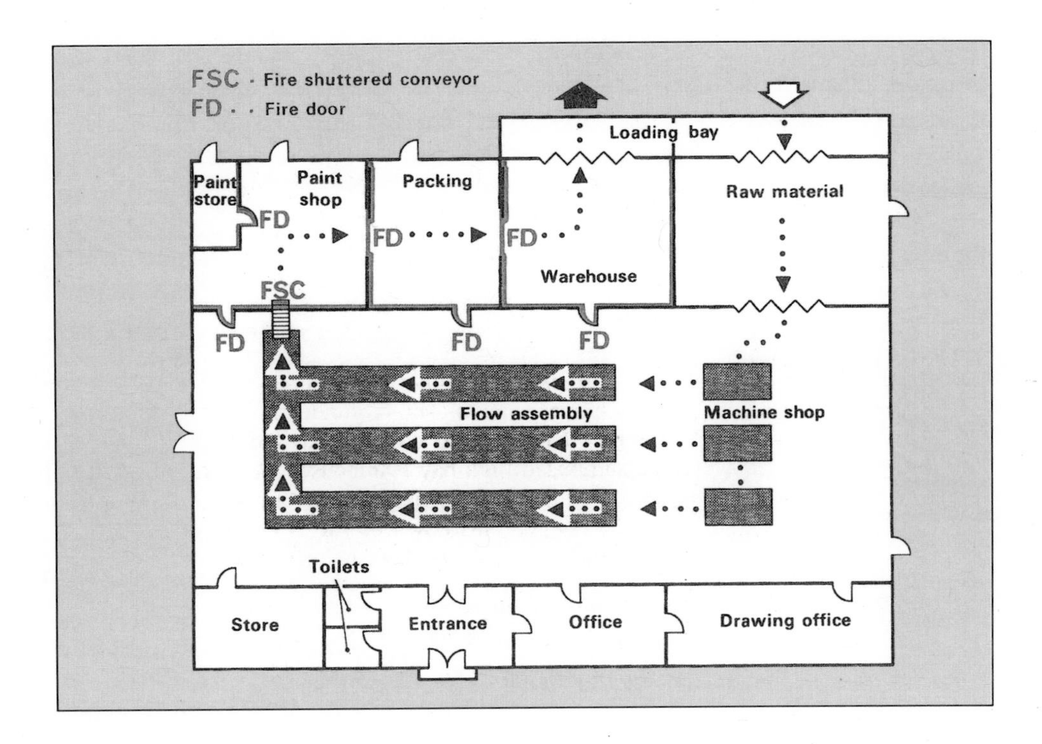

Figure 7.14 A simple diagram of a factory layout.

Automatic **H**igh-**B**ay **W**arehouses (AHBW) or **H**igh-**B**ay **W**arehouses (HBW) These vary in size and degree of automation but the latest are almost completely computer controlled which cuts down on the labour force and enables stock levels at retail outlets to be kept to the minimum. AHBW are constructed in two basic ways:

(a) An integrated structure where the racking constitutes the support for the roof and the wall cladding is attached to it. Heights of 30m are common, and
(b) Where the racking is separate from, and does not support, the walls and roof. These tend to be smaller probably up to about 15m high.

The floor areas and cubic capacity are large. Some dimensions and a brief description of one such building is as follows:

High bay building size – 155m × 55m × 22m
Cubic capacity – 187,550m^3
Floor area – 8,525m^2
Total pallet storage – 11 high × 24 wide × 112 long = 29,568 pallets
No. of cranes – 6 = 4,928 pallets/crane, fully automatic, computer driven
Throughput – Up to 3 pallets/min or 2 pallets in and 2 out per min, simultaneously
Mass of racking and stock – 40,000t
Air conditioning Temperature – 13°Celsius at 3% humidity.

The building is Portal framed with walls comprising 50mm steel/isocyanurate/steel composite cladding and a very low-pitched roof of similar insulated material.

7.17.2 Low-bay area

Apart from the cranes the whole stock is moved on computer controlled conveyors to and from road and rail unloading and dispatching bays, which cover another 3,500m^2 of low bay floor area.

7.17.3 Fire protection

Automatic protection in this particular building is of a high standard. There are a large number of smoke and rate-of-rise detectors arranged in zones. The main indicator board for the system is in the works fire brigade station.

There are 6 large smoke vents at each end of the HBW and 9 of the roof trussing are enclosed in steel sheeting to act as smoke curtains.

Apart from hose-reels and extinguishers, the main firefighting system is a specially designed sprinkler system. The sprinklers are fitted both in the roof and in-rack and are so arranged that, for example, if smoke or fire is detected at level 2, sprinklers at levels 2, 3, 4 and 5 will operate. The system is backed up by 3 pumps, 2 of, which are diesel driven, and 2 very large tanks of water.

Figures 7.15 and 7.16 show external and internal views of this complex.

7.17.4 Firefighting

As each HBW is different in construction, access, automatic protection, water supplies, numbers of people, plans passed by the local Fire Authority will differ considerably.

If the planning of the site has been done correctly the means of escape for persons on the premises will be adequate according to the numbers likely to become involved. People could, however, still fail to evacuate in time and become missing. Searching an automatic HBW can be hazardous. Figure 7.16 shows the complication at the "front end" of the racks. Depending on the system used, this can be even more complicated between racks with cables, conveyor belts, picking cranes, maintenance platforms etc., which, although stopped from operating, still present hazards.

Visits to the premises by firefighters are essential so that they have some idea of the environment they could be working in, possibly in smoke. Most AHBW have a preponderance of smoke vents but this does not rule out the possibility of smokelogging under the worst conditions. BA control at an incident would have to be of a very high standard.

7.17.5 Contents of a High-Bay Warehouse

The contents are so varied in the usual type of HBW supplying a supermarket or hypermarket chain that a policy of the separation of hazardous goods is usually in operation. However, there are chemical HBWs, which although being extra well protected, always present the possibility of accidents, spillage, explosions and fires. Obviously the brigade and the management will have carried out preplanning but, according to the type of incident, decisions will have to be made on such points as:

(a) Whether to commit BA personnel into the building if there are no people involved. This will depend somewhat on the type of construction. Early collapse could be expected if a really hot fire ensued;
(b) Chemical protection and subsequent decontamination;
(c) Type of extinguishing media to be used;
(d) Damage control – when to begin;
(e) Evacuation of people down-wind if a dangerous chemical cloud or plume of toxic smoke is emitted from the incident.

Figure 7.15 A high-bay warehouse 155m × 55m × 22m, totally enclosed. The six large smoke outlets are repeated at the other end.

Photo: Rowntrees Mackintosh.

Figure 7.16 The interior of a high-bay warehouse showing automatic picking cranes and the extremely restricted conditions in the racking area.
Photo: Rowntrees Mackintosh.

7.18 Public assembly buildings and Town Centres

7.18.1 Public assembly buildings

General

This group comprises, amongst others, theatres, cinemas, concert halls, leisure centres, museums, art galleries, churches, schools, non-residential clubs and bingo halls.

Many of these premises will be subject to some form of licence usually issued by the local authority. Such licences generally impose terms, conditions and restrictions relative to fire safety. In most cases the local authority is required to consult with the fire authority before issuing or refusing a licence. Buildings in this group are constructed in many forms and examples of typical assembly buildings are described.

7.18.2 Theatres

The traditional theatre consists of a substantial outer wall and, internally, the seating area or auditorium is divided into separate part floors at varying levels e.g., circles, balcony, gallery. Each of these has sufficient exits leading to protected staircases and to the safety of open air at ground level.

Between the auditorium and the stage are the proscenium openings. Scenery, in the form of painted fabrics, plastics, various types of curtains and other large pieces of material are stored high over the stage, some on rollers in the hanging loft or "fly gallery". This is so named because such scenic effects are termed "flown" or "flying" scenery. Also in the stage area are the electronic and electrical controls for lighting and sound, and other set properties ("props") which are moved into the stage area from the scene dock as required.

The fire risk in the stage area is considerable so the stage area is isolated from the auditorium by a substantial wall called the proscenium wall. Only the minimum openings are allowed in this wall one of which is the stage itself. A fire-resisting curtain, known as the "iron" protects this, and this is, itself protected by a curtain drencher system (see Figure 7.17). Roof ventilators, in the form of haystack lantern lights, are installed over the stage area for the rapid automatic clearance of smoke.

Many modern theatres/concert halls do not use flying scenery and their stages are not separated from the auditorium. These are called "open stage" or "theatre-in-the-round". Stage sets in this type of theatre are inherently of low combustibility or durably flameproofed. The potential fire risk is a lot less than that encountered in the traditional "picture-frame" type of theatre. Methods of presentation are being continuously developed and the design of theatres and the conditions of licence under which they operate are adapted to the varying methods of production (see Figures 7.18, 7.19 and 7.20).

Irrespective of the design of the theatre, the safety of the audience is still the paramount consideration.

7.18.3 Cinemas

The older type of cinema, of which there are still a few, were constructed in a similar way to theatres

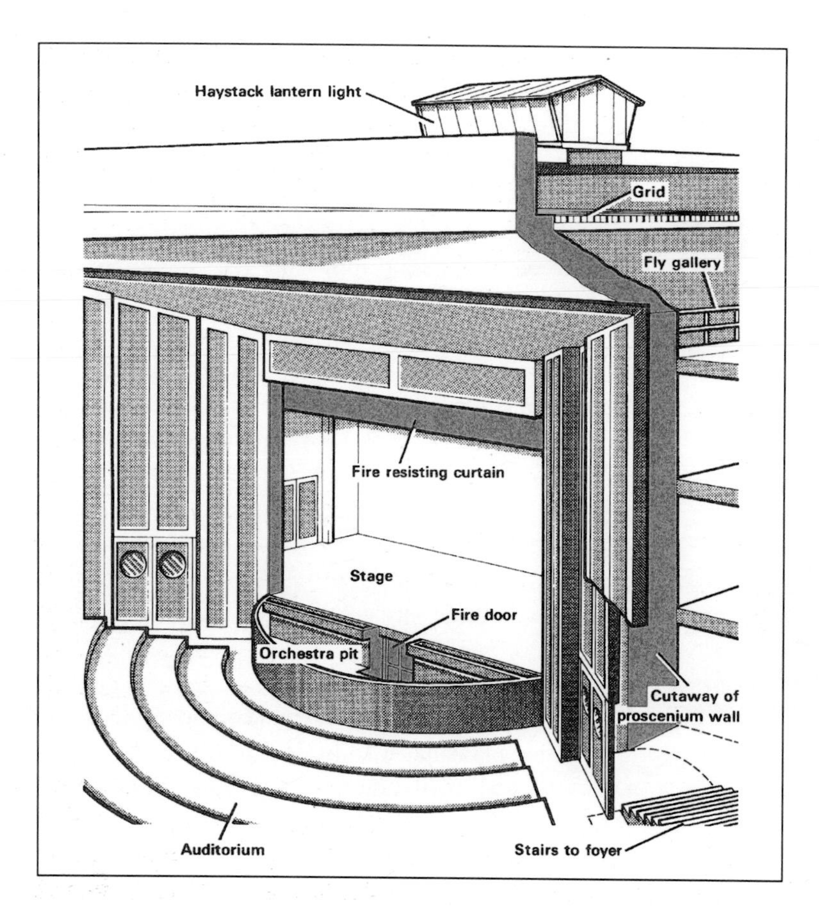

Figure 7.17 Diagram of a traditional small stage theatre showing separation between auditorium and stage area.

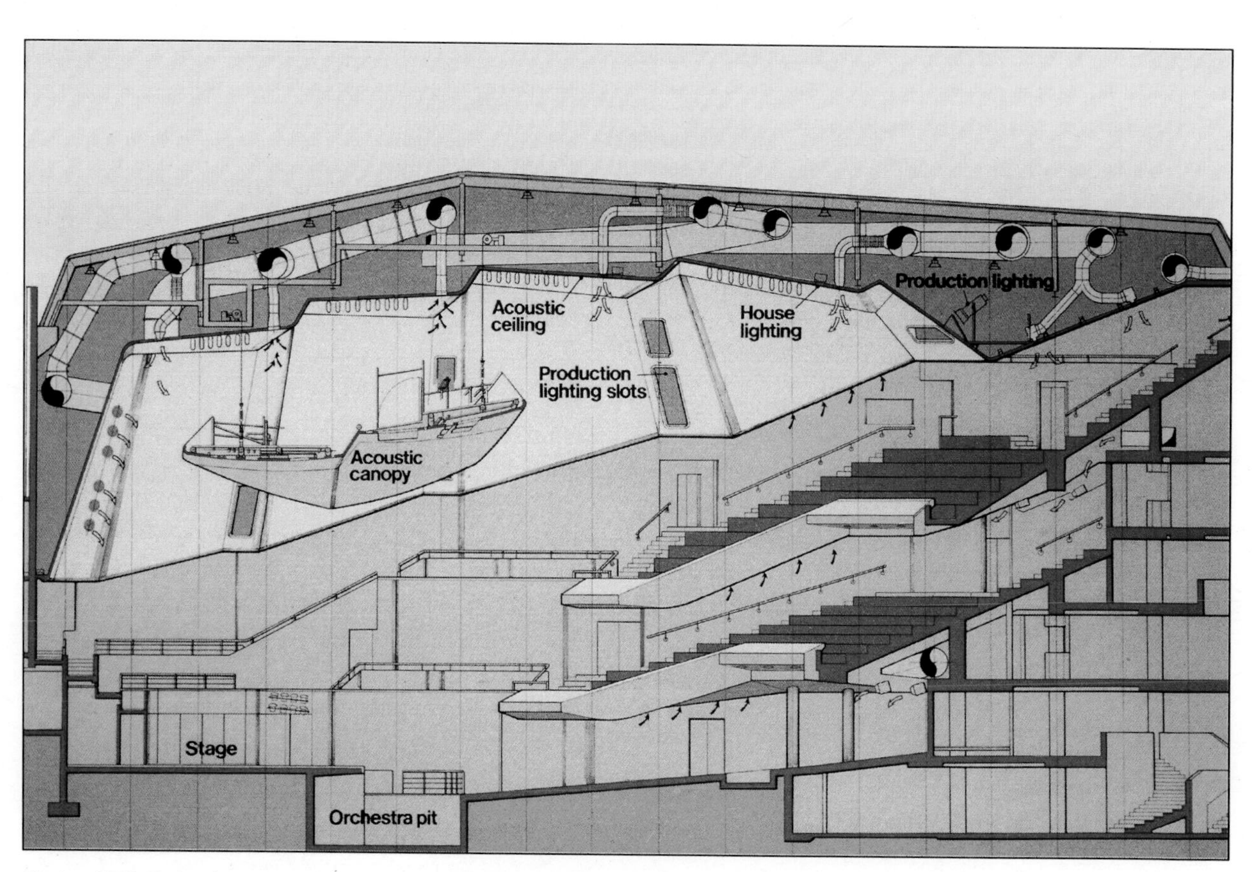

Figure 7.18 Side elevation of a typical modern theatre/concert hall. Photo: RHWL Partnership.

Figure 7.19 Interior of the theatre/concert hall illustrated in Figure 7.18. Photo: RHWL Partnership.

and controlled in a similar way. The trend, nowadays, is for the multi-auditoria cinema and this has raised problems.

If the cinema is a purpose-built single storey, multi-auditoria type then the means of escape from all auditoria is usually straight to open air.

It is the conversion of the old single auditorium cinemas to multi-auditoria that bring the problems. Dependent on the way the cinema is sub-divided e.g., horizontally or vertically or, perhaps, both, it is often found that the means of escape would be a problem for one, or more, of the auditoria. Perhaps the number of exits is sufficient but the travel distances are excessive or too many exits converge into one means of escape e.g., one foyer.

Obviously these problems have to be resolved to the satisfaction of the licensing authority but firefighters need to know their way about the rather more complicated conversions and be aware of circumstances which may arise should an evacuation be necessary.

7.18.4 Sports and leisure centres

This type of complex is being built all over the UK in increasing numbers either as a separate entity or as part of a town centre, theme park etc., Sizes vary from what amounts to a swimming pool plus a few squash or badminton courts through to complexes which include facilities for practically every indoor sport and outdoor sport plus restaurants, bars, children's playgrounds, roofed stadia, lecture/concert halls etc.

The numbers of people using these complexes at any one time can be high and include all ages. It would be difficult to restrict numbers and,

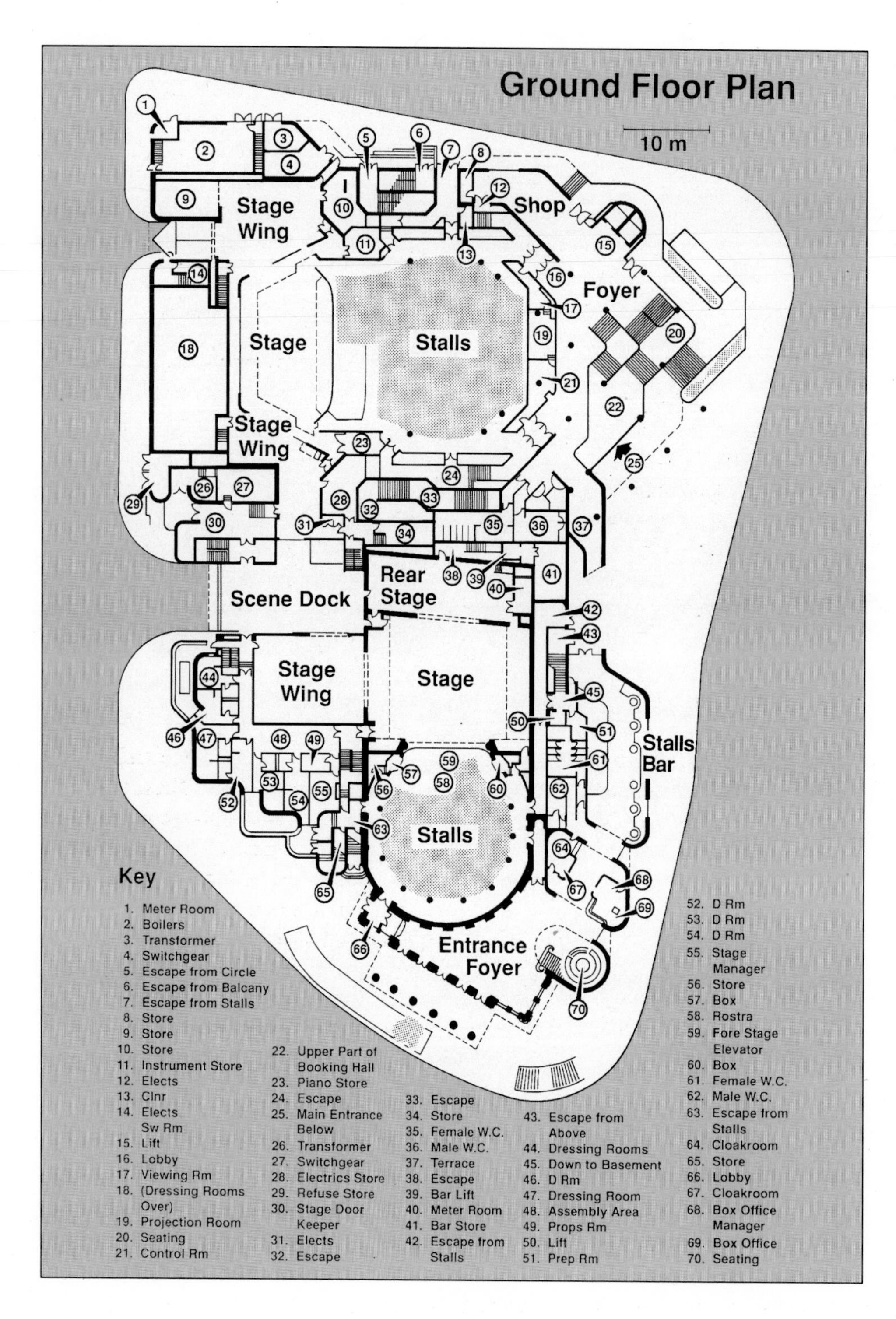

Figure 7.20 The ground floor plan of combined traditional and modern theatres under one roof.

therefore, great care is taken to ensure adequate fire precautions, protection and means of escape.

Construction and design are individual to a site but, being new buildings, all are subject to Building Regulations. Various pieces of legislation will apply depending on the actual use of the various parts of the complex but the "Guide to the fire precautions in existing places of entertainment and like premises, 1990", although having no statutory force does direct firefighters to the relevant Acts, Codes of Practice, Guides and British Standards. Two particular aspects are mentioned:

(a) The surface finishes and furnishings on escape routes;
(b) The use of cellular foams e.g., in gymnastic mats, and their safe storage.

British Standards and the Furniture and Furnishings (Fire) (Safety) Regulations 1988 are relevant but are just among a large number of guidance documents which could apply. There is also the "Guide to safety at sports grounds, 1990" which advises, amongst other things, on the fire safety aspects regarding crowds of people.

7.18.5 Town Centre development

This type of building development is frequently called a "complex" and, although it may include a variety of multilevel, multi-occupancy buildings, the whole area is regarded as an entity. It can consist of a completely new environment or the conversion of an already existing area by roofing over, pedestrianisation, enclosure etc.

Some of the largest developments will occupy several acres and include a theatre/cinema/concert hall, department stores, supermarkets, open markets, leisure centres and offices. Podiums, escalators, bridges, galleries, staircases may connect these, entire wall-climbing lifts etc., and they almost always include atria. British Standard 5588 Part 7 "Atrium Buildings" provides basic guidance on the problems. Examples are shown in Plates in the Manual, Book 9.

With the prospect of the presence of large numbers of people of all ages and abilities, the means of escape is of paramount importance. Other aspects requiring thought will be access for brigade appliances, water supplies, smoke venting, sprinklers. Modern developments of this sort are examples of "intelligent" complexes with a control room overseeing a large number of electronic devices, which monitor, report and warn staff of any adverse conditions, which might arise.

As with all new or materially altered buildings, these developments have to be constructed in accordance with the Building Regulations. To make a commercially viable yet reasonably safe environment architects, planners, building control officers and fire prevention officers have to work closely together.

BS 5588 Pt. 10 "Enclosed shopping complexes" deals with the subject in considerable depth.

Each development is individual and many different styles of building and building materials are used in their construction. Where possible, firefighters should take the opportunity to inspect the construction site as it progresses so that have a good knowledge of not only what will be eventually seen but much of what will be concealed.

7.19 Air-supported structures

7.19.1 Types of Air-supported structures

An air-supported or pneumatic building, is generally a type of structure which consists of a single membrane anchored to the ground and kept in tension by internal air pressure so that it can support applied loading (see Figure 7.21). A main fan maintains the internal air pressure or fans with provision for automatically operated standby fans. There are variations of this type of system, for example:

(a) An air-inflated structure in which air is contained within ribs formed of a membrane of PVC or other plastics or fabric. These ribs form the "structural elements": the columns and beams which in turn support the roof and walls.
(b) An inflated double walled structure in which air is contained between the membranes.
(c) An air-supported structure, which consists of a single membrane, supported by a small pressure above atmosphere (inflation) over the whole of the structure's internal surface.

Particular designs have different characteristics from the point of view of occupant safety; for example, in the case of (c) the designer will often incorporate lightweight steel framing for the primary purpose of providing a stable fixing for lighting equipment, but this may serve also to support the roof should the structure deflate for any reason.

7.19.2 Use

Air-supported structures are used in many areas because of the low capital cost and minimum maintenance costs. They are used for commercial and industrial applications, such as warehousing and manufacturing processes, for the agricultural

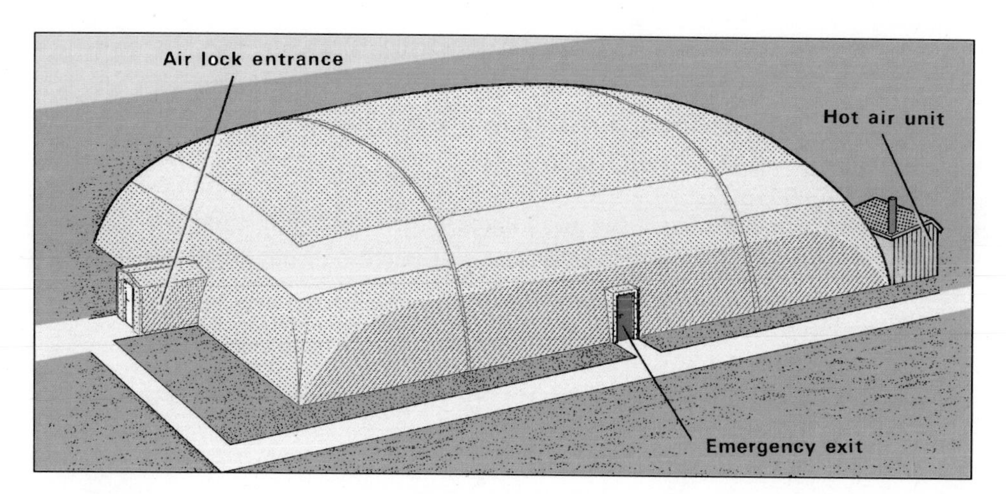

Figure 7.21 An example of an air-supported structure.

and horticultural industries and for military use a s garages, radar equipment protection, mobile hospitals and workshops. They will also be found as enclosures for swimming pools, tennis courts and other sports, which are affected by adverse weather conditions.

7.19.3 Behaviour of structure in fire

The membrane, itself, can be made from either a coated fabric i.e., nylon or polyester with a PVC or rubber coating or, alternatively, from plastic, PVC or polythene sheet material. BS 6661 refers to the behaviour of membranes in the event of a fire and states that the membrane should not readily support combustion. Experience has shown that PVC coated polyesters, polyamides and unreinforced polythene generally perform satisfactorily under fire conditions, melting rather than burning in a fire. Any holes formed by such melting would allow the membrane to sag with possible collapse onto the fire. Many factors are involved including the type of fabric used, the nature of jointing substance, the height and extent of the structure, and the size and type of fire.

7.19.4 Means of escape

The fundamental requirement for the safety of the public in the event of fire breaking out in any of these structures is to make sure that they quickly become aware of the danger and are able to reach a place of safety before being overcome by smoke, toxic gas or other products of fire. It is imperative therefore that every structure is provided with exits and emergency exits. BS 6661 gives guidance on the number of exits, their width, siting and travel distances. Ironically the door most suited to the smaller air-supported structure is the revolving door as this causes the least air loss. However, as it takes so much longer for a given number of people to pass through this type of door as compared to a conventional type, revolving doors are usually considered to be unsuitable from an escape point of view. When they are permitted, an additional door or opening must be provided, bringing more problems from air loss. Air locks and air curtains are recognised means of access and exit, but they must conform to BS 6661 requirements. No exit width should be less than 800mm and, where the occupancy exceeds 100 persons this should be increased to 1100mm with extra width for every extra 100 persons. Travel distances are restricted and emergency lighting and exit signs conforming to BS 5266 Pt 1 are required.

7.20 Underground and unfenestrated buildings

7.20.1 General

If involved in fire these types of premises are especially dangerous to both occupants and fire fighters. Whether they are large sub-basements, systems of tunnels and caverns or above ground totally enclosed buildings, the problems include:

(a) An absence of adequate means of venting the heat, smoke and toxic products of combustion;
(b) A general lack of access and subsequent chance of disorientation;
(c) Difficulty in appraising the fire conditions or

even finding the fire without risking firefighters in dangerous areas;

(d) Communications between firefighters and between them and the outside;
(e) Application of extinguishing media;
(f) Congestion and restriction of movement within the space involved.

A CFBAC working party studied the subject and published its findings in FSC 4/1968. Their comments fell mainly into two groups – operational and fire prevention. Many of the points made have become standard practice in brigades e.g., guide lines, communications and BA procedures.

7.20.2 Operational

Any fire in this type of premises brings extremely punishing conditions of excessive heat and humidity. This may require large numbers of BA wearers in order to keep the exposure of individual firefighters down to a tolerable level. Very strict control of BA procedure must be maintained. It is to be expected that any brigade having this type of premises in their area would pre-plan and hold regular exercises and/or visits so that firefighters likely to become involved have some knowledge of the risk. This knowledge could include:

(a) Difficult areas of communication;
(b) Alternative means of access;
(c) Water supplies;
(d) Position of internal doors, shutters and lifts;
(e) Probable location of BA controls.

Some of the premises could be high security e.g., MOD. Any difficulties in obtaining access or knowledge would be subject to negotiation. These must, obviously be carried out with the emphasis on the fact that the lives of LAFB firefighters are at risk.

7.20.3 Fire prevention

Generally, purpose-built premises of this sort would be compartmented, automatically protected and vented and organised to control hazardous storage. However, most of these premises are adaptations and the 1968 report recognised this and made recommendations accordingly. Amongst these were the following:

(a) Effective control of the type of contents and isolation of the particularly hazardous elements;
(b) If staff are present, adequate training in the use of the firefighting equipment e.g., hose-reels, and extinguishers and how to raise the alarm;
(c) The installation of automatic detection and adequate means of calling the brigade;
(d) Compartmentation, fire-stopping etc., to be maintained to a high standard;
(e) Adequate means of escape for the occupants;
(f) Plans of access etc., well-indicated and available for firefighters;
(g) Where necessary, the installation of fixed automatic firefighting systems e.g., sprinklers, foam, halons, powder;

7.21 Restoration/Refurbishment and Re-use of old buildings

The rehabilitation and re-use of old buildings has, during the past two decades, become a major component of construction activity and currently accounts for a large proportion of the buildings industry's work load.

Building owners and developers have come to realise the potential value of our vast stocks of old buildings as a means of providing modern accommodation more quickly and at a lower cost.

There has also been a significant increase in attitudes favouring conservation rather than the wholesale demolition and redevelopment policies of the 1960s which resulted in large numbers of attractive, substantial old buildings being replaced with "modern" buildings of a lower standard.

The rehabilitation and re-use of old buildings is, like new construction, subject to statutory constraints, and the need to comply with various forms of legislation can have a significant effect on the viability of such schemes.

Complying with current demands of fire safety standards can involve a considerable amount of work and expense in upgrading the structure, providing means of escape and fulfilling compartmentation requirements, possibly making an otherwise attractive scheme economically prohibitive.

Old buildings may already be subjected to other legislation. They may be "Listed" buildings as being of special architectural or historic interest. As such, it is an offence to carry out works of complete or partial demolition, alteration or extension in any manner which would affect its character without having first obtained listed building consent from the local planning authority.

The number of listed buildings in Great Britain is now about half a million. It is quite likely therefore, especially in our older cities and towns, that a building being developed might be listed. However, it is possible in the majority of cases to carry out sensitive rehabilitation and alterations of listed buildings and fire officers may find themselves increasingly being asked for advice in this respect.

Legislation requires every local planning authority to determine which parts of its area are areas of special architectural or historic interest, and to designate those areas as conservation areas.

Many buildings, whilst they may not be listed, stand in conversation areas and as an essential part of the conversation area's character or appearance, the development possibilities may be limited.

The rehabilitation and re-use of old buildings presents a challenge to the fire safety officer in terms of sympathetic consideration of fire safety aspects.

Guidance is also available as follows:

The Construction Unit, Department of the Environment.

Historic building conservation guide for government departments. A guide to the care and use of historic buildings in England, Wales and Northern Ireland.

The guide describes the policies, principles and procedures to be observed and gives guidance in doing so. It is not intended as a technical book though reference is made to technical matters.

Similar guidance is available in Scotland: Historic Scotland 1997. Technical Advice Note 11 'Fire Protection measures in Scottish Historic Buildings'.

Chapter 8 – Services in buildings

8.1 Introduction

Most modern buildings are designed to include what are known as services, e.g. air-conditioning, heating lifts, dust extraction plants, electrical circuits.

These will usually require roof or floor air-conditioning plants, boiler rooms, lift motor rooms, fans, and electrical transformer rooms and, perhaps, access floors. Often very large buildings or building complexes are, as has already been stated, intelligent; i.e. all services are centred on a control room, which is usually manned although much of the equipment in the building will be self-monitoring. This section deals with natural and mechanical ventilation, dust and solvent extraction, conveyors and machinery drives.

8.2 Natural and mechanical ventilation

8.2.1 Natural ventilation

In natural ventilation, the circulation and renewal of the air inside a building is effected by a combination of air entering from outside the building and air currents generated inside. Air enters on the windward side through doors, windows and ventilators and is drawn out by suction on the leeward side and up chimneys even though there is no fire.

Natural ventilation is greatly increased by the internal air currents set up by heating (radiators etc.) and the natural warmth of the occupants. The warmed air rises and escapes through the tops of windows and high level ventilators, and cold air is drawn in to replace it through the doors and windows.

Many different types of air inlets and outlets have been used at different periods and in different kinds of buildings. They are all, in essence, simply holes in the outer wall or roof fitted with flaps, grilles or louvers to allow air to enter or escape and, at the same time, excluding rain and draughts. The type most commonly used today is known as an "air-brick" and is placed at high level as an outlet in rooms, which have no chimney flue. It consists of a 230mm square opening in the wall protected on the inside and outside by grilles. Similarly air bricks are provided to ventilate the underside of a wooden ground floor where fitted, the air entering on the windward side, circulating round the house and out on the other side.

Low-level air inlets are uncommon today but many old building have them, often screened by a metal tube which deflects the air upward (see Figure 8.1). In many modern buildings, particularly offices and hotels, fresh air inlets are often provided behind radiators placed under windows.

Older single storey factories and large halls are often ventilated by means of lantern lights in a flat roof or louvered ventilators placed at the apex of a pitched roof. More modern factories may have automatic ventilators or louvers fitted some of which are rain sensitive. These can all provide useful means of controlling smoke in a building but could tend to draw fire into a roof space.

8.2.2 Mechanical ventilation

In mechanical ventilation the circulation of air is assisted, or even carried out entirely, by a system of fans and ducting. It is used in many buildings such as warehouses, cinemas, theatres, offices etc. where large numbers of people congregate in a relatively small space. It is also used in factories where the occupants must be protected from harmful atmospheres or dust produced during a manufacturing process.

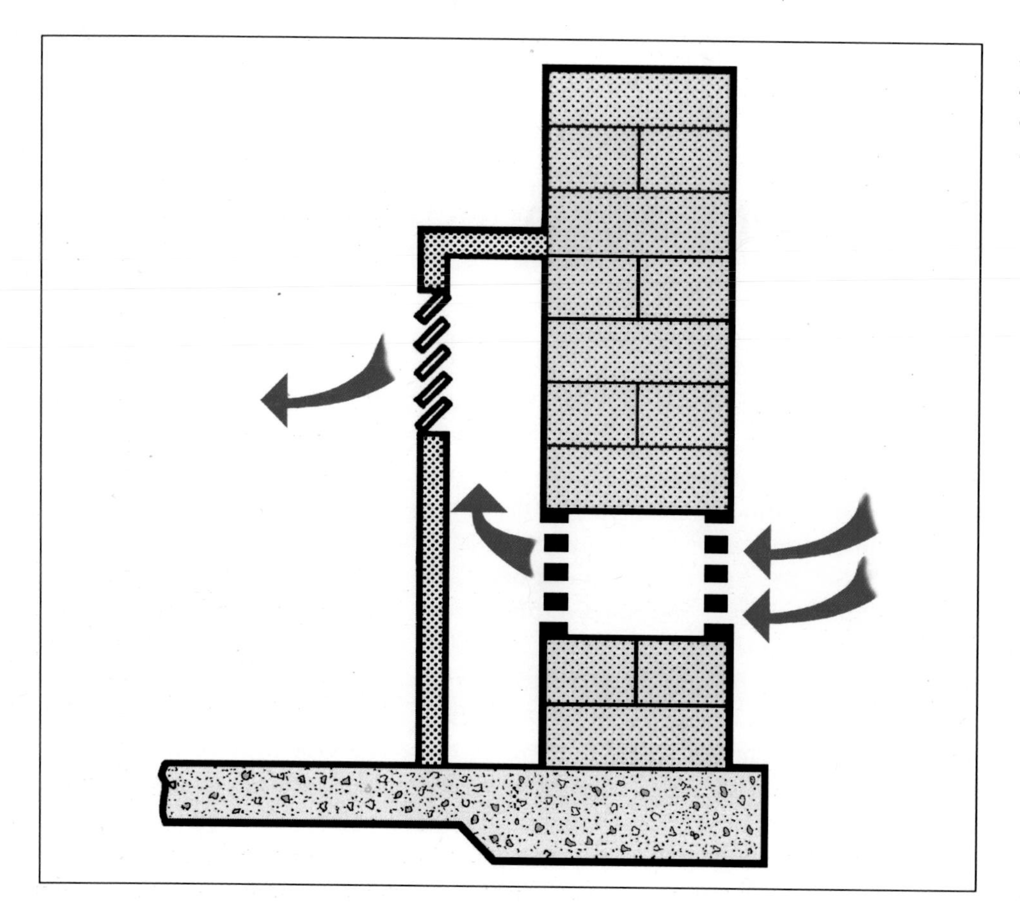

Figure 8.1 Diagram showing the arrangement of a low level inlet.

Mechanical ventilation can be divided into three principal groups:

(a) Where the stale (vitiated) air is extracted from the building by fans, fresh air finding its way in through doors and windows;
(b) Where fresh air is forced into the building by fans, stale (vitiated) air finding its way out through doors and windows;
(c) Where fans are used both to force fresh air into the building and to drive out stale (vitiated) air.

In the last two methods the air pressure inside the building is kept slightly above that outside so as to avoid incoming draughts i.e. they are balanced systems. All mechanical systems include ducting, usually made from steel sheeting, which distributes or extracts the air.

Firefighters must be aware of the possibility of spread of fire or smoke throughout the building via these systems. Figure 8.2 shows, diagrammatically, an arrangement of the conditioning plant and parts of the ducting of a balanced system in large premises.

8.2.3 Air conditioning systems

These systems are, in effect, extensions of ventilating systems in that they provide ventilation air, which has been warmed or cooled and has the desired level of humidity. Basically an air conditioning system consists of:

(i) Fans for moving the air;
(ii) Filters for air cleansing;
(iii) Refrigerating plant connected to heat exchange surfaces such as finned coils or chilled water sprays;
(iv) Means for warming the air;
(v) Means of humidifying the air;
(vi) A control system to regulate the amount of heating or cooling automatically.

Depending on the type of use of the building some modern buildings (including healthcare premises) have zoned systems (each zone having its own

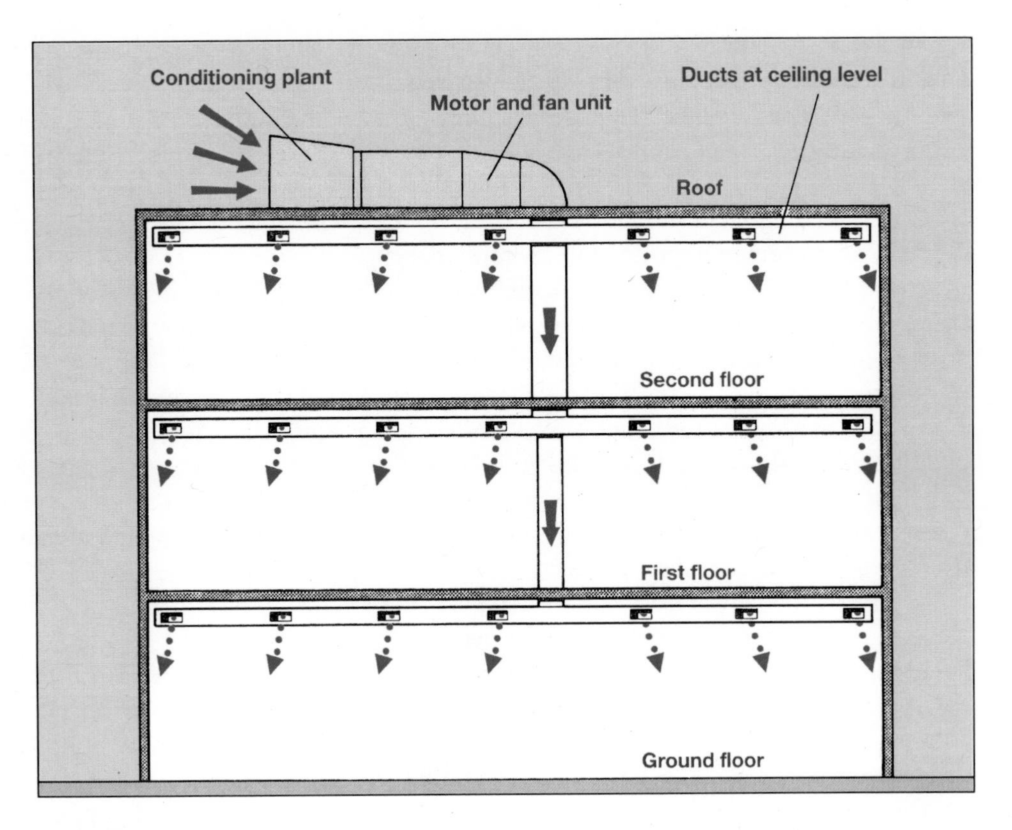

Figure 8.2 Diagram showing the arrangement of a "plenum" system for an industrial premises.

minisystem) but more usually the top floor is given over to the air-conditioning plant and the cleaned, heated/cooled, humidified air is circulated round the building from there. On their own these systems present little hazard and most systems will have automatic controls which operate whenever a fire situation is monitored or can be manually controlled from the central control room. Smoke is then either prevented from spreading around the building or, if required by the brigade, extracted.

8.2.4 Dust extraction

This is usually required by legislation in commercial or industrial premises to either protect the occupants from excessive inhalation of dust or the removal of a possible explosive atmosphere hazard. Finely divided dust, of almost any kind, has the potential of extremely rapid flame propagation or explosion once an ignition source is introduced. Dust is usually collected, by extractor fans, through ducting to a collecting area. Any spark at the machine end can quickly be drawn along a duct and, as soon as it reaches the correct mixture of oxygen and dust can cause an explosion. Quite often a minor explosion happens first which disturbs more dust and this larger cloud can cause a massive detonation. Any fire in a duct is dangerous and firefighters should attempt to cut off the forced draught first and then operate fixed sprinklers or introduce sprays into the duct.

8.2.5 Fume extraction plants

In practically all processes involving chemicals, especially liquids, e.g. paint spraying, there will be a need to extract the fumes in order to keep processes safe. In some cases the fumes may be recovered, processed themselves and stored or recirculated into the system. Here again automatic fire protection is usually required and safety regulations are strict. Fires do occur in the ducting and consequently, these are sometimes fitted with pressure relief panels either to vent an explosion or to enable access to be made to the duct. Firefighters should take care and follow a similar procedure, as above, not forgetting to cover any storage area.

8.3 Mechanical conveyors and chutes

8.3.1 Conveyor belts

Nowadays, most conveyor belts, and similar automatic methods of transporting goods, are highly sophisticated and are found not only in factories and warehouses but also in shopping complexes and hypermarkets. Because they are constructed mainly with non-combustible materials they do not, generally, present a fire hazard.

Existing building control also expects a high degree of automatic protection where they penetrate fire-resisting walls and floors. Modern conveyors run on specially designed rollers which do not require grease or oil, which eliminates one of the most prevalent causes of fires on belts. There is still the possibility of static electricity building up and, unless the earthing is sound, providing a source of ignition with sparks. Some older heavy-duty conveyors are still run on metal rollers and guides and the old hazard of grease build-up is still there. Firefighters should try to stop a conveyor if called to a fire on or near one because this prevents burning material being carried around a building. Occasionally fixed automatic sprinklers are fitted over the conveyor and these should be left operating until other firefighting media can be brought into use. An example of mass-transfer of goods is the high-bay warehouse described elsewhere and others are in such industrial areas as vehicle manufacturers. Though not all strictly "belts" they are all practically continuous mass movement systems with similar problems of maintenance and possible overheating of motors etc.

8.3.2 Gravity feeds

Gravity feeds will be found either in the form of pipes or channel slides. Both forms are used in factories and warehouses. Piping is, of course, used for conveying industrial liquids from one part of a factory to another usually from tanks on upper floors or the roof. Such liquids as acids, alkalis and flammables used in industrial processes may be moved in this way. Obviously any leakage can lead to a very dangerous situation and firefighters should, wearing appropriate protective clothing, attempt to isolate the fracture by operating valves.

8.3.3 Suction pipes

Another method of mass transfer is by suction piping. Such materials as cement, grain, sugar, pulverised fuel, plastic pellets etc. are sucked out of the bulk transport e.g. a ship's hold and fed, via piping, to their storage areas. Grain elevators and fuel silos are often filled in this way. As with any smallgrained material the danger is of a dust explosion and there are various systems preventing either the dust from attaining the right explosive mixture or of venting the pressure if it does explode.

Firefighters must take great care when working in, near or over any of these types of storage area. The use of BA would prevent dust from damaging the lungs but, more importantly, would provide air if a firefighter fell and was buried by a fall of the material. The use of lines attached to firefighters would also make it easier to trace anybody who fell into the material and to haul them out.

8.3.4 Machinery drives

Most machinery in commercial and industrial use is driven by electrical power. The feeding of the power varies from light 3-core cable, running a cloth cutter or tailor's iron to a heavy multi-core mineral insulated copper sheathed (MICS) cable powering a heavy crane. The design of the system is always peculiar to the task required to be done even to the point of requiring intrinsic safety e.g. for explosive atmospheres.

Any fires involving this type of power-drive will usually be due to the overheating of the electrical motor. Most systems will have either an automatic cut-off for the power i.e. an overload switch or a type of manual "punch" button to isolate the machine rapidly.

In large workshops containing a lot of machinery firefighters must move about with caution as machinery may still be running.

8.4 Heating systems

Heating systems in buildings, other than private houses, are by warm air or hot water systems. These are usually run from a boiler fired by gas, oil, solid fuel or, very occasionally, electricity. The systems are such that, in addition to usually being enclosed in a fire-resisting compartment, automatic safety devices are fitted which, in the event of dangerous conditions arising, will cut off the fuel supply and convey an alarm.